HOW TO IMPLEMENT
WASTE-TO-ENERGY PROJECTS

HOW TO IMPLEMENT
WASTE-TO-ENERGY
PROJECTS

by

Marc J. Rogoff, Ph.D.

np NOYES PUBLICATIONS
Park Ridge, New Jersey, U.S.A.

363.728
R73h

Copyright © 1987 by Marc J. Rogoff
No part of this book may be reproduced in any form without permission in writing from the Publisher.
Library of Congress Catalog Card Number: 87-12211
ISBN: 0-8155-1132-9
Printed in the United States

Published in the United States of America by
Noyes Publications
Mill Road, Park Ridge, New Jersey 07656

10 9 8 7 6 5 4 3 2 1

Library of Congress Cataloging-in-Publication Data

Rogoff, Marc Jay
  How to implement waste-to-energy projects.

  Includes bibliographies and index.
  1. Refuse as fuel--United States. 2. Waste products as fuel--United States. I. Title.
  TD796.2.R64 1987   363.7'28   87-12211
  ISBN 0-8155-1132-9

To
Phyllis and Leslie

Preface

A waste-to-energy project is perhaps the most complex public works project usually attempted by a community. For successful implementation, it needs the services of knowledgeable in-house staff and consultants and requires the active support of the community's political establishment. Oftentimes, the decisions involving the siting, permitting, and financing of such projects are controversial and are opposed by vocal groups of citizens.

Why then should communities undertake such projects in the wake of such obstacles? The answer lies in the fact that many communities in the United States have no other answer to the problem of the ever mounting increase in their solid waste disposal needs. Restrictions on the siting and operations of sanitary landfills makes waste-to-energy a feasible alternative.

This book is designed for use by those responsible for solid waste disposal in America's communities. Its purpose is to present sufficient information on what steps must be undertaken to implement a waste-to-energy project. While every project is unique to some extent, my hope is that the lessons learned on already implemented waste-to-energy projects can lead to more successful projects in the future.

Tampa, Florida
May, 1987

Marc J. Rogoff, Ph.D.

About the Author

Marc J. Rogoff, Ph.D., is a Project Manager for HDR Techserv, Inc., located in their Tampa, Florida office. He is responsible for the planning and development of solid waste and waste-to-energy/resource recovery projects throughout the United States. He has experience for both government and industry in the implementation of waste-to-energy projects. He previously served as Hillsborough County, Florida's Resource Recovery Program Administrator and his efforts were instrumental in the implementation of this project, now in operation. Dr. Rogoff has a Ph.D. from Michigan State University, an M.B.A. from the University of Tampa, and M.S. and B.S. degrees from Cornell University. He serves as Chairman of the Resource Recovery Committee of the Governmental Refuse Collection and Disposal Association, and is an active member of the Technical Steering Committee of the Coalition on Resource Recovery and the Environment (CORRE), the National Resource Recovery Association, and the National Solid Waste Management Association.

Acronyms and Abbreviations

A/E	architect/engineer
BACT	Best Available Control Technology
Btu	British thermal unit(s)
C	degree(s) Celsius
CFR	Code of Federal Regulations
CO	carbon monoxide
CO_2	carbon dioxide
cty	county
DER	Department of Environmental Regulation
DOI	Department of Interior
e.g.	for example (exempli gratia)
EPA	U.S. Environmental Protection Agency
ESP	electrostatic precipitator
et al.	and others (et alia)
F	degree(s) Fahrenheit
FAA	Federal Aviation Administration
FERC	Federal Energy Regulatory Commission
ft	foot (feet)
G.O.	general obligation
gr/dscf	grain(s) per dry standard cubic foot
HHV	higher heating value
H_2S	hydrogen sulfide
IDB	industrial development bond
i.e.	that is (id est)
IFB	Invitation for Bid
kwh	kilowatt-hour(s)
LAER	Lowest Achievable Emission Rate
lb	pound
msw	municipal solid waste

Acronyms and Abbreviations

mg/l	milligrams per liter
Mw	megawatt
NA	not available
NAAQS	National Ambient Air Quality Standards
NIMBY	Not-In-My-Back-Yard
NPDES	National Pollutant Discharge Elimination System
No.	number
NO_x	oxides of nitrogen
NSRSA	New Source Review Standards Analysis
p.	page
PCDD	polychlorinated dibenzodioxin
PCDF	polychlorinated dibenzofuran
pp.	pages
PSC	Public Service Commission
PSD	Prevention of Significant Deterioration
psi	pounds per square inch
PURPA	Public Utilities Regulatory Policies Act
QF	qualifying facility
RDF	refuse derived fuel
RFP	Request for Proposal
RFQ	Request for Qualification
scf	standard cubic foot (feet)
SIP	State Implementation Plan
SO_2	sulfur dioxide
t	ton(s)
TCDD	tetrachlorinated dibenzodioxin
tpd	ton(s) per day
$\mu g/m^3$	microgram(s) per cubic meter
U.S.	United States

NOTICE

To the best of the Publisher's knowledge the information contained in this publication is accurate; however, the Publisher assumes no responsibility nor liability for errors or any consequences arising from the use of the information contained herein. Final determination of the suitability of any information, procedure, or product for use contemplated by any user, and the manner of that use, is the sole responsibility of the user.

The book is intended for informational purposes only. The reader is warned that caution must always be exercised when dealing with materials or procedures which might be hazardous.

Contents

1. **INTRODUCTION AND OVERVIEW** 1
   The Solid Waste Disposal Problem 1
   The Trends Towards Waste-to-Energy 2
   References 6

2. **PROJECT IMPLEMENTATION CONCEPTS** 7
   Introduction 7
   Developing the Project Team 10
      Internal Project Team 11
      Consultants and Advisors 12
   Risk Assessment 14
      Waste Stream 16
      Energy and Materials Market 17
      Legal and Regulatory 17
      Facility Construction 18
      Facility Operation 19
   Implementation Process 19
      Project Phases 20
         Phase I—Feasibility Analysis 20
         Phase II—Procurement 22
         Phase III—Plant Construction 22
         Phase IV—Plant Operations 23
      Implementation Project Scheduling 23
      Implementation Project Costs 25
   Public Information Programs 25
   References 29

3. **WASTE-TO-ENERGY TECHNOLOGY** 31
   Introduction 31

Mass Burning 33
    Process Description 35
    Operational Experience 39
Modular Combustion 42
    Process System 42
    Operating Facilities 43
Refuse Derived Fuel (RDF) Systems 46
    Processing Systems 47
        Wet RDF Processing 47
        Dry Processing Systems 48
Fluidized Bed Systems 54
Anaerobic Digestion 56
Composting 56
Pyrolysis Conversion Systems 58
Comparison of Technologies 58
    Energy Efficiency 58
    Cost 60
References 61

4. SOLID WASTE COMPOSITION AND QUANTITIES 64
Introduction 64
Characterization of Solid Waste 65
Waste Composition Studies 67
Heating Value 69
Solid Waste Quantities 71
References 72

5. WASTE FLOW CONTROL 74
Introduction 74
Flow Control Mechanisms 76
    Waste Flow Control Through Legislation/Regulation 76
    Contractual Control of Waste Stream 80
    Economic Incentives for Waste Stream Control 81
References 82

6. SELECTING THE FACILITY SITE 84
Introduction 84
The Site Selection Process 85
    Evaluation Criteria 85
    Technical Considerations 88
        Site Drainage 88
        Foundation Suitability 88
        Size and Shape of Site 89
        Accessibility 90
        Location 90
        Utilities 92
    Environmental Considerations 93
        Air Quality 93

      Water Quality 94
      Biological Resources 95
    Social Considerations............................... 95
      Surrounding Land Uses........................... 95
      Permitting Considerations........................ 96
      Land Ownership 97
      Cultural Resources.............................. 97
    Site Screening Process 97
      Stage 1: Data Collection and Analysis.............. 98
      Stage 2: Preparation of Constraint Maps........... 98
      Stage 3: Identifying Potential Site Areas 100
      Stage 4: Preliminary Screening of Site Areas 100
      Stage 5: Evaluating and Selecting Candidate Sites...101
      Stage 6: Site Selection..........................104
        Land Use Compatibility 105
        Environmental Impact 105
        Comparative Costs..........................107
    References...108

7. ENERGY AND MATERIALS MARKETS......................110
    Introduction......................................110
    Energy Markets....................................111
      Steam 111
      Electricity 113
    Materials Markets 117
      Ferrous Metals 119
      Non-Ferrous Metals............................119
      Glass...120
      Paper Products 121
      Film Plastics...................................121
    References..122

8. PERMITTING OF WASTE-TO-ENERGY FACILITIES 123
    Introduction......................................123
    Federal Permits....................................124
      Air Quality....................................124
      Water Quality..................................133
      Federal Aviation Administration Permit...........135
    State and Local Permitting Requirements...............136
    Keys for a Successful Permitting Strategy..............138
    The Dioxin Issue 139
    References..141

9. PROCUREMENT OF WASTE-TO-ENERGY SYSTEMS............145
    Introduction......................................145
    Procurement Approaches 146
      Architect/Engineer (A/E) Approach 148
      Turnkey Approach 149

xiv Contents

  Full-Service Approach 150
 Procedures for Conducting the Procurement Process 150
  Competitive Sealed Bidding............................. 151
  Multiple-Step or Simultaneous Negotiations Method 153
  Competitive Negotiation................................ 154
  Sole-Source Negotiation 156
 Preparing the Request for Proposal 157
  Format of an RFP..................................... 158
  General Information to Proposers....................... 159
  Instructions for Proposal Preparation and Submission......... 160
  Technical Requirements 163
  Proposal Forms....................................... 164
  Draft Agreements 164
  Evaluation, Selection and Negotiation Process.............. 164
 Proposal Evaluation 165
  Log-In Procedure and Proposal Handling 165
  Evaluation Committee 166
  Review of Completeness and Conformance with RFP......... 167
  Detailed Evaluation 167
 Negotiations Process 169
 References... 171

10. OWNERSHIP AND FINANCING OF A WASTE-TO-ENERGY FACILITY.. 173
 Introduction... 173
 Ownership Alternatives.................................. 175
  Public Ownership 175
  Private Ownership 175
 Financing Options...................................... 178
  General Obligation (G.O.) Bonds 178
  Project Revenue Bonds................................. 179
  Grant Funds, Loan Guarantees, and Entitlements 182
  Private Equity.. 183
 Key Participants in Resource Recovery Financings.............. 184
  Bond Counsel.. 184
  Independent Consulting Engineer........................ 185
  Investment Banker.................................... 186
  Financial Advisor 187
  Trustee 187
  Rating Agencies 188
 Steps in Bringing the Bond Issue to Market.................... 189
  Step 1. Adoption of a Bond Resolution and Trust Indenture.... 189
  Step 2. Validate the Bonds............................. 190
  Step 3. Preparation of a Preliminary Official Statement 191
  Step 4. Meetings with Rating Agencies and Bond Insurance Firms... 193
  Step 5. Blue Sky and Legal Investment Surveys 193
  Step 6. Establishing the Final Pricing of the Issue 194

　　　　Step 7. Submission of the Purchase Contract to the Issuer 195
　　　　Step 8. Bond Issue Closing 196
　　　　Step 9. Post Sale Activities 196
　　References 197

INDEX 199

- 1 -
Introduction and Overview

THE SOLID WASTE DISPOSAL PROBLEM

How to dispose of the cans, cereal boxes, newspapers, tires, bottles, and other castoffs of America's communities in an environmentally sound and economically efficient way has become a problem of critical proportions. Counties and cities across the United States are being confronted with the ever increasing volume of solid waste generated by our society's residential, commercial and industrial activities. It has been estimated that Americans currently generate over 250 million tons of solid waste annually, or an average of five pounds per person per day on a nationwide basis. With population growth and waste generation rates expected to increase upward in the years ahead, many communities are beginning to search for alternative long-term solutions to the methods they now employ to dispose of their solid wastes.

Sanitary landfilling of solid waste has become the traditional approach for most communities. Landfilling has progressed from an earlier era of dumps and open burning to its present state. Sanitary landfills can be designed today to be an environmentally acceptable means of waste disposal, provided they are properly operated. New regulations regarding landfill liners, leachate control systems, landfill gas collection and control systems, and long-term closure requirements, however, have dramatically increased the cost of landfilling. In addition, suitable land for landfill sites close to nearby urbanizing areas is now less available for many communities, thereby resulting in communities having to locate more distant landfills. The Not-In-My-Back-Yard (NIMBY) attitude on the part of citizen opposition groups, however, has increased the difficulty of many communities in the siting and permitting of these new landfills. Consequently, as the nation's existing landfill capacity has been reduced, there has been an increased interest in the concept of recovering energy and recyclable materials from municipal solid waste rather than relying on sanitary landfilling as the primary long-term method of solid waste disposal.

THE TRENDS TOWARDS WASTE-TO-ENERGY

Producing and utilizing energy from the combustion of solid waste is a concept which has been practiced in Europe since the turn of the last century. Prompted by a concern for groundwater quality and the scarcity of land for landfilling, many European countries and Japan embarked on massive construction programs for waste-to-energy programs in the 1960's. Transfer of this technology to the United States first began in the late 1960's and early 1970's. In addition, many other projects utilizing American technology in the area of shredded and prepared fuels were constructed. Most of these projects were dismal failures, however, because they were unable to overcome materials handling and boiler operations problems. It was these failures that made local government leaders cautious in funding construction of waste-to-energy projects.

Nevertheless, several waste-to-energy projects were developed in the mid to late 1970's in communities such as Saugus, Massachusetts; Pinellas County, Florida; and Ames, Iowa which were experiencing severe landfill problems. Success of these projects helped the waste-to-energy industry gain acceptance by local government leaders, and the financial community. Tax incentives made available by the federal government for waste-to-energy projects attracted private capital investment in such projects assisting in the maturing of this industry and sparked the development of many new projects. In the period of 1984-85, for example, approximately $6.5 billion worth of waste-to-energy bonds were sold to finance construction of these facilities.

This frenzied trend in waste-to-energy project financings slowed in early 1986 due to uncertainties over the impact of the new tax laws enacted by Congress, and by a steep reduction in energy revenues in many parts of the nation. The long-term need for environmentally sound alternatives for solid waste disposal, however, will continue to drive many communities to evaluate the feasibility of constructing waste-to-energy facilities to satisfy these needs.

While less than five percent of our nation's solid waste is currently being processed in waste-to-energy facilities, Europe and Japan long ago surpassed the United States in utilizing waste-to-energy technology for disposal of their municipal solid wastes. Denmark, for example, currently converts 70% of its solid waste to energy; Switzerland 40%; Sweden 50%; and Japan 40%. According to a survey completed by the National Resource Recovery Association in 1986, there were only 69 waste-to-energy facilities in operation in the United States disposing of about 24,000 tons of solid waste per day (Table 1-1). However, some 29 plants are currently in the design or construction stage. When these plants are completed by the early 1990's, waste-to-energy facilities will be processing from 15 to 18 percent of all the municipal solid waste being produced in this country.

Industry market surveys completed within recent years suggest that development of waste-to-energy facilities will be concentrated in those areas of the nation that are experiencing difficulties in siting new sanitary landfills, where population growth is rapidly increasing, and where strict environmental regulations for landfills have been enacted. The plants constructed in these areas, however, will generally be smaller than the plants in operation or under construction today. This is due to the fact that most of the communities, who could support utilization of large waste-to-energy plants in the range of 1,500 to 3,000 tons per day, would have already implemented these projects. Mid-sized plants (i.e., 500 to 1,500 tons per day) and smaller scale plants (i.e., less than

TABLE 1-1
STATUS OF U.S. WASTE-TO-ENERGY FACILITIES, 1986

Technology Classification	Operating	Closed	Status Construction	Planned
Mass Burn	17	4	16	61
Modular	36	14	6	16
Prepared Fuel (RDF)	16	12	7	14

Sources: References 1, 2, and 4.

500 tons per day) will provide enormous market opportunities, which some have estimated to be in the range of $15 to $20 billion till the year 2000.

REFERENCES

1. Berenyi, Eileen and Robert Gould, A Progress Report. *Waste Age* August: 48-53 (1986).

2. U.S. Conference of Mayors, Report on Semi-Annual Survey: Resource Recovery Activities. *City Currents*, April (1986).

3. Walter, Donald K., Energy and Waste: A Status Report. *Waste Age* May: 97-102 (1986).

4. Waste Age Magazine, The Waste Age Refuse-to-Energy Guide. *Waste Age* November: 197-212 (1986).

- 2 -

Project Implementation Concepts

INTRODUCTION

The successful implementation of a waste-to-energy project rests primarily upon the following essential building blocks or key elements:

- o A reason or need for the project because of a critical community solid waste disposal problem or crisis;

- o An implementing government agency or private project developer with political commitment willing and able to undertake the project;

- o An adequate supply of solid waste for the project or means to assure waste stream control or attract sufficient quantities from other communities;

o Markets for the recovered energy and recovered materials; and

o A project site that is environmentally, technically, socially, and politically acceptable.

Perhaps the most critical element that must be in place if a waste-to-energy project is to succeed is that a need for the project exists. That is, a situation exists such that community leaders perceive that the community is facing an immediate or long-term solid waste disposal problem, and that planning for an alternative to sanitary landfilling should be undertaken.

A second major element that must be present for the success of a waste-to-energy project is political leadership. Unfortunately, the most well- conceived plans for public benefit projects often are left unimplemented without such leadership. Waste-to-energy projects are capital intensive and require planning that can often extend over two to five years time. Since most politicians are elected every two or four years, waste-to-energy projects can find themselves orphaned by new political leaders who may have different solid waste management agendas. Consequently, if a community has any hope of implementing a waste-to-energy facility, then it is necessary to have a implementing entity (e.g., a county, municipality, authority, electric utility, waste hauler or contractor) or driving force which has long-term political support.

The community must also be able to supply or divert enough solid waste to its proposed facility since solid waste serves as the feedstock for plant operations and energy production. The community must be able to both guarantee the quantity and quality of its solid waste. The level of the wastes to be guaranteed will determine the ultimate size of the facility.

Securing an energy and materials market is another critical component in implementing a waste-to-energy project. Such markets provide revenues that offset plant tipping fees and make waste-to-energy facilities financially attractive to both communities and private developers. Absent these markets, waste-to-energy projects would not prove economically feasible for most communities.

Another critical project component in project implementation is securing a site to construct and operate a facility. Siting public benefit projects, like solid waste facilities, has proven to be a time-consuming and controversial task in recent years. Such projects have attracted significant public opposition because of citizen concerns associated with perceived project impacts such as: air quality, public health, traffic congestion, litter, noise, aesthetics, and property values. The Not-In-My-Backyard (NIMBY) attitude has caused project developers to spend more time in searching for project sites that are technically, environmentally, and socially acceptable.

What is somewhat unique about waste-to-energy projects, aside from their complexity, is the tortuous paths they often have taken from project inception through construction. This lengthy implementation period for many projects has resulted in part from their inability to assemble an experienced project team having strong, long-term political support to move the project along when faced by major project impediments. Absent this strong management organization and support, many projects have been delayed by their inability to resolve such critical problems as: strong public opposition to a proposed site, addressing environmental concerns, and unfavorable project economics. While chapters three through ten will concentrate on the critical contractual, technical, and financial decisions that each project must successfully resolve, this chapter will primarily focus attention on developing the management structure in order to cope with these project issues on a day-to-day basis and the means to effectively communicating these complex issues to the public and the people overseeing the project.

DEVELOPING THE PROJECT TEAM

An extremely important aspect of assuring success of a program with the complexities of a waste-to-energy project is establishment of a strong project team that can guide such a long-term project to completion. While every project is unique to some extent, there are a number of fundamental actions which government must be prepared to take to place a project on the right course for the long-term. Such measures can enhance the project's potential for success.

Internal Project Team

At the outset of a project, a key action is the establishment of an internal team which will have support from the political decision-makers. Since such projects will require significant up-front development costs for staff and consulting services over several years, it is critical that there be a long-term commitment by the community to support the project. Without this commitment, it is unlikely that the project will truly ever succeed.

Ideally, the internal project team, which will direct the activities of the government agency's own staff and outside consultants or advisors, should have appropriate agency heads from their key administrative, public works, financial, legal, environmental, and communications areas. The purpose of this interagency committee is to guide the project through key decision points and to provide policy recommendations to the political decision-makers. Since the agency heads of all key government departments are members of this project committee, it is more likely that project decisions will receive a more balanced and thorough review before presentation to the community's elected representatives, and result in unified staff recommendations. One disadvantage of this type of project structure, however, is that it requires extensive time commitments on a continual basis from government departments, which may be unable, because of their own project demands and budgets, to supply these services. Thus, many projects have been unable to organize such an interagency management committee in practice.

To compensate for a lack of full-time support from outside agencies, many solid waste, public works, planning, or utility departments responsible for implementation of their community's waste-to-energy projects have established the position of a full-time project manager. They have recognized that the coordination of their government's staff and outside consultants and advisors was essential to the success of the waste-to-energy project. The person selected for this position often comes from within the particular government department having responsibility for the project, or is hired from outside government as a contract employee. Regardless of this individual's civil service status, his or her role typically is to be responsible to coordinate, schedule, and monitor the activities of the internal project team and consultant staff. This individual is often assisted by other full-time staff members because of the tremendous significant time required to successfully undertake these roles. As an example, the Hillsborough County, Florida Resource Recovery Project had a full-time project coordinator who was assisted by another full-time staff member during its feasibility and implementation phases.

Consultants and Advisors

A consulting team with an excellent track record in waste-to-energy implementation should be hired at the outset of the project to complement the government's internal project team. Many governments hesitate to utilize consultants at early phases in such projects because of the significant costs associated with obtaining consulting services. Unfortunately, this is shortsighted because consultants are usually cost effective in the long-run since they will add credibility and needed expertise to a community's project.

An independent consulting engineering firm, which has significant experience in the waste-to-energy area, can provide the community with valuable insights. The feasibility report which it prepares can point out the advantages and disadvantages to the community with respect to such key issues as: project ownership, financing, procurement, siting, and permitting. If nothing else, its assistance in drafting the project's request-for-proposal should enhance the quality of this procurement document which can result in more responsive vendor proposals to the community's needs.

Since the bond issues for most waste-to-energy projects are usually the largest ever issued by most local governments, it is also important that a strong outside financial team be selected to prepare the overall financing for the project. This team usually includes one or more bond underwriting or investment banking firms, an independent financial advisor, a bond counsel, and a bond underwriter's counsel. Each of these firms or individuals has a specific role in helping develop a strong financing plan for the project which will be favorably viewed by bond rating agencies and the credit markets. Chapter ten describes some of these different roles in detail.

As the project nears the procurement phase, other experts are sometimes added by some governments to their consulting team. For example, an insurance advisor is sometimes hired to develop the technical insurance requirements for the Request-for-Proposal, and to assist in securing insurance coverage for the project. In addition, special legal counsels are often retained to assist in environmental permitting; help negotiate energy sales contracts; or assist in the preparation and negotiation of construction and operations contracts with the selected contractor for the facility.

RISK ASSESSMENT

An assessment of the possibility that a single or multiple event will occur, which will have a detrimental impact upon a project, and who should bear this loss, are key components of the feasibility analysis for a waste-to-energy project. Many of the decisions that a community makes along the path of project implementation are concerned with allocation or assignment of risk events in the following general areas: waste stream control; energy markets; legal and regulatory arrangements; project construction; and project operation. While it is difficult to determine the probability of each particular risk or exposure, which usually results in the monetary loss, such events can be categorized by who was responsible for the cause of the event (i.e., the community, contractor, or force majeure). Table 2-1 lists the major risk events in a waste-to-energy project and the typical assignment among project participants.

Risk sharing among project participants is usually achieved through the negotiation of contracts. While any party can assume a project risk, it is more typical in waste-to-energy projects

TABLE 2-1
TYPICAL ASSIGNMENT OF PROJECT RISKS

Risk Event	Procurement Approaches			
	A/E	Turnkey	Full Service Public	Full Service Private
Waste Stream:				
Failures in Waste Stream	Government	Government	Government	Government
Changes in Waste Composition	Government	Government	Government	Government
Energy and Materials Markets:				
Energy and Materials Revenues	Government	Government	Government	Government
Legal and Regulatory:				
Tax Law Changes	Government	Government	Government	Shared
Environmental Permitting	Government	Government	Government	Shared
Anti-Trust Challenges	Government	Government	Government	Government
Facility Construction:				
Design Errors	Government	Contractor	Contractor	Contractor
Equipment Performance	Contractor	Contractor	Contractor	Contractor
Costs Underestimated	Shared	Contractor	Contractor	Contractor
Failure of Subcontractor	Shared	Contractor	Contractor	Contractor
Subsurface Conditions	Government	Government	Government	Government
Inflation	Either	Either	Either	Either
Strikes	Either	Either	Either	Either
Force Majeure	Government	Government	Government	Government
Facility Operation:				
Plant Performance	Government	Government	Contractor	Contractor
Damage by Waste	Government	Government	Government	Government
High Inflation	Government	Government	Government	Government
Operating Costs	Government	Government	Contractor	Contractor
Residue Disposal	Government	Government	Government	Government

Source: Reference 8

for the party responsible for controlling the cause of a potential loss to be allocated that risk. Assumption of risk beyond the reasonable control of a party usually requires that he be adequately compensated for such risk sharing. Thus, while full-service contractors can be expected to assume greater project risk than contractors either in an A/E or turnkey procurement approach, there is usually a direct correlation between the level of such risk and the overall construction price of the waste-to-energy project. This is due to the nature of the project performance and corporate financial guarantees that a full-service contractor must assume. Consequently, it is critical that a risk assessment be made early in a project so that certain potential project events are understood and a risk and compensation posture is developed.

Waste Stream

Historically, risks associated with assuring that a reliable supply of solid waste is delivered to a waste-to-energy facility are assumed by the community. This means a community must be able to fully guarantee that solid waste generated within its boundaries will be delivered to its proposed waste-to-energy facility. Waste stream control can be achieved through a number of methods such as: the use of state legislation or local ordinance; long-term put-or-pay contracts; subsidized tipping fees; or governmental collection of solid waste.

In addition to guaranteeing delivery of solid waste to a waste-to-energy facility, the community usually must assume the risk of the quality of such waste. That is, the heating value or Btu content of the waste, its percentage of moisture, and percentage of combustibles. The community typically guarantees a reference solid waste composition which a contractor assumes for design purposes. Thus, if this reference composition changes, perhaps due to increased recycling of paper products, then the community must assume the responsibilities and costs of these changes, generally through an increase in tipping fees.

Energy and Materials Market

The risks associated with energy and materials markets are typically assumed by a community. In the event of lower prices for energy and recycled products than anticipated, the community is usually responsible for subsidizing project revenues through increased tipping fees. These risks may be partially mitigated by securing negotiated long-term contracts with the energy or materials customers.

Legal and Regulatory

Risks associated with changes in laws and governmental regulations, which are generally unforeseen or uncontrollable, are generally allocated to a community. It is difficult at the outset of a project to anticipate changes in critical federal or state legislation and rules in areas such as tax law, environmental protection issues, and antitrust challenges in the courts. Through negotiation with a full-service contractor, which desires to own the community's waste-to-energy facility, these tax-law risks can often be allocated to the contractor.

Facility Construction

Risks associated with construction of a waste-to-energy facility include the following: design errors; strikes; failure of subcontractors to perform; equipment performance; cost underestimated; subsurface conditions; inflation; and force majeure. Typically, those risks associated with the performance of the design, technology, or equipment are the contractor's responsibility. Furthermore, risks within the contractor's reasonable control are typically his financial responsibility. Contractors for waste-to-energy facilities will generally guarantee construction price (subject to normal inflation), the length of construction, and the performance of his subcontractors to be on time and within budget. These guarantees are generally supported through construction and performance bonds, project insurance policies and incentives by a community to encourage timely completion. The exception is in the case of the A/E procurement approach where these risks are typically shared with the community through negotiation. For example, a maximum contract price could be assigned to a contractor under the "chute-to-stack" procurement approach.

Another cause of risks associated with project construction are force majeure or uncontrollable events. These risks can include acts of G-D, such as floods, earthquakes or other natural disasters; sabotage; war; explosions; and unforeseen site conditions, such as subsurface soil conditions. Historically, these risks are assumed by government which can mitigate these losses by procuring special insurance for the project.

Facility Operation

Risks associated with the operation of a waste-to-energy facility include: continued plant performance; damage to the plant by the community's solid waste; inflation; operating costs; and residue disposal. Under both the A/E and turnkey procurement approach, a community assumes all the risks of continued performance and operation costs of the project after its warranty period. In contrast, the community has the option to share some of these risks or pass them on to a contractor under the full-service procurement approach. Historically, full-service operators will guarantee specified plant performance levels and maximum annual operating and maintenance expenses. These guarantees are sometimes subject, however, to occurances beyond the control of the contractor such as: above normal inflation; damage to the plant by hazardous or explosive materials contained in the solid waste stream; or inability to dispose of ash residues to a legally permitted landfill. Such risks are generally assumed by the community.

IMPLEMENTATION PROCESS

Implementation of a waste-to-energy project is a complex process which consists of several phases requiring "go/no-go" decisions to be made by the project participants. Since no two waste-to-energy projects in the United States are identical, the discussion on project implementation in this section is by nature generic in approach. Table 2-2 displays the steps normally followed in the implementation process for a typical waste-to-energy project. These steps can be modified by a community taking into account its individual concerns and needs. The critical point is that a well-planned implementation process is essential to an expeditious, cost-effective, and successful project.

Project Phases

Phase I - Feasibility Analysis. This phase is preceded by a preliminary, and somewhat informal, investigation by a community or project sponsor that the major project elements, as discussed above, exist, and make further study of a potential waste-to-energy project reasonable.

During Phase I, the feasibility of such a project is evaluated in detail. An analysis is undertaken of the community's existing and projected waste stream to determine the ultimate size of a single or multiple unit, waste-to-energy system. Markets are examined to determine whether the energy and/or materials produced by this system can generate adequate revenues to offset the construction, operating, and financing costs for the facility. Feasible sites are investigated along with an analysis of the technical, environmental, and institutional requirements for permitting a facility. In addition, a risk assessment is undertaken to help determine the risk posture to be taken by the community with regards to waste-to-energy technology, project ownership, operation, and financing. A formal feasibility report, which is presented to

TABLE 2-2
TYPICAL IMPLEMENTATION PHASES OF A WASTE-TO-ENERGY PROJECT

Phase I - Feasibility Analysis

o Waste Stream Analysis   o Review Permitting Requirements
o Waste Disposal Practices Analysis   o Risk and Legal Assessment
o Energy and Materials Market Study   o Financial Analysis
o Analysis of Feasible Waste-to-Energy Technologies o Develop Project Alternatives
o Analysis of Potential Facility Sites   o Go/No-Go Decision

Phase II - Procurement

o Select Project Alternative   o Develop Financing Plan
o Select Site and Acquire   o RFQ/RFP Produced and Issued
o Permitting Underway   o Contractor Selected
o Market Contracts Concluded   o Contract Negotiations Concluded
o Waste Stream Guarantee   o Notice-To-Proceed

Phase III - Plant Construction

o Site Preparation   o Equipment Installed
o Complete Final Design   o Testing and Startup
o Equipment Ordered   o Acceptance Testing
o Building Constructed   o Certificate of Completion

Phase IV - Plant Operations

o Service Fee Payment   o Annual Report (Optional)
o Annual Tipping Fee Adjustment   o Facility Retesting (Optional)

the implementing entity, usually documents all activities in this phase. Phase I ends with a "go/no-go" decision on the part of the community to either proceed with implementation of the project or to terminate activities for the foreseeable future.

Phase II - Procurement. Phase II incorporates all the steps necessary to procure the waste-to-energy system desired by the community including: contracts for waste supply, energy and materials markets, plant construction and operation (if applicable); acquiring the project site; obtaining all environmental permits and/or regulatory approvals; and securing the financing for the project. Phase II usually requires the development of specialized procurement documents such as a Request-for-Qualifications (RFQ) and Request-for-Proposal (RFP). Dependent on the financing approach selected by the community, documents such as a Bond or Trust Indenture, Engineer's Feasibility Report, and Bond Prospectus may need to be written.

Phase III - Plant Construction. Phase III covers all the steps necessary in the construction of the waste-to-energy facility including: site preparation; final design; ordering facility equipment; installing this equipment; testing and startup of the facility; and completing acceptance tests. These tasks are usually undertaken by the selected facility contractor, although most communities obtain the services of a knowledgeable engineering consulting firm to independently monitor construction activities and the facility's acceptance tests. Phase III usually ends with the facility successfully completing the contractual acceptance test procedures and the community issuing a certificate of completion to the contractor.

Phase IV - Plant Operations. Phase IV covers the period from plant acceptance through the length of the operations period, which may last 20 years or more in the case of a full-service contractor for a publicly or privately- owned facility. These contracts usually have clauses allowing them to be renegotiated after this time period. In contrast, operations contracts for turnkey operators can be significantly shorter, generally five years or less.

During this period of plant operations, communities may have a varying degree of project responsibility. In the case of governmental operations, they have the full range of operational responsibilities. In the case of a privately-owned and operated facility, however, the community may only have the right to request that the facility be retested to prove that its original performance guarantees can be met. Most communities retain independent consulting engineering firms to undertake these oversight roles.

Implementation Project Scheduling

The duration of the implementation process for a waste-to-energy project depends upon a variety of factors which project sponsors may or may not control. Factors such as procurement approach, delays in regulatory reviews, and length of negotiations for contracts, for example, impact the length of time which a community may need to conclude the implementation of a waste-to-energy project. Consequently, it is difficult to provide a rule of thumb for project duration which will apply to all projects. Generally, however, a full-service procurement approach in waste-to-energy projects can be completed faster than either the architect/engineer (A/E) or turnkey approaches. This is primarily because the community can simultaneously undertake all the remaining procurement activities on a fast-track basis with limited project delays.

With significant numbers of waste-to-energy projects being implemented in recent years, it is expected that the length of time required to implement such projects can be reduced. Early projects required upwards of two to three years to complete planning and procurement activities, with an additional two years to complete all contractual negotiations and arrange for financing. This time period was in addition to the two to three years required for design and construction of the facility. Thus, many communities desiring to implement a waste-to-energy project had to look forward to an extended period of time before their facility would be fully implemented.

Current experience suggests that waste-to-energy projects can be implemented in a somewhat shorter period of time. It appears that as feasibility procedures become more refined, Phase I activities can be completed in an average of six months time followed by Phase II procurement activities lasting an additional 12 to 18 months. Once contracts are concluded, and financing arranged, design and construction activities can be completed in an average of 24 to 36 months. Some projects have even further reduced these schedules by fast-tracking procurement, financing, and construction activities.

Implementation Project Costs

Because they are highly capital intensive, waste-to-energy projects require significant amounts of upfront expenditures on the part of sponsors for implementation activities. The following activities on the part of staff and consultants often require expenditures depending on the complexity of the waste-to-energy project: consulting engineering services; legal, financial and insurance advice; land acquisition; subsurface soil testing; land surveys; payment and performance bonds; municipal bond insurance; liability and force majeure insurance; bond ratings and credit reports; printing of official statements and bonds; miscellaneous bond issuance expenses; and expert witnesses. Table 2-3 is a listing of typical expenses required to implement selected waste-to-energy projects, now in operation or in construction. These project initiation expenses are generally recovered by communities through issuance of long-term bonds.

PUBLIC INFORMATION PROGRAMS

An effective public information program is essential to the success of any waste-to-energy project. The public information program can be, in many respects, similar to that used in most public benefit projects. For example, details concerning the project can be disseminated to the general public through the following typical media: periodic

TABLE 2-3
IMPLEMENTATION COSTS ($ MILLIONS) OF SELECTED WASTE-TO-ENERGY PROJECTS

Facility Location	Year	Bond Size	Development Costs	Issuance Costs	Equity Contribution	Bond Discount
Tampa, FL	1984	115.6	2.3	0.7	2.0	3.5
Baltimore, MD	1983	190.8	1.0	5.5	63.1	5.0
Commerce, CA	1984	44.2	---	0.3	5.0	1.6
Lawrence, MA	1982	58.2	2.2	8.9	30.8	3.3
Hillsborough County, FL	1984	144.0	5.3	3.5	3.6	3.8
Hennepin County, MN	1986	129.3	21.3	5.4	23.1	d
Indianapolis, IN	1985	109.0	---	1.2	22.2	1.7
Norfolk, VA	1984	107.8	2.7	1.1	---	2.7
Olmsted County, MN	1985	25.0	0.2	0.1	6.2	0.4
Pinellas County, FL	1980	160.0	2.5	1.8	---	3.4
Saugus, MA	1975	30.0	0.2	1.2	10.0	---
Tulsa, OK	1984	58.5	3.8	2.8	15.9	1.6
Westchester County, NY	1982	157.4	---	4.9	79.5	4.7

Notes:

a Development costs include site acquisition and consultant's fees related to implementation.
b Issuance costs include legal fees, bond insurance premium, printing fees and certain other costs associated with financing.
c Equity contributions include state grants, community and private contributions.
d Bond discount included in cost of issuance.

project newsletters; brochures; press releases; movies; slide shows; visual displays; press packets; radio and television interviews by project and staff consultants; presentations before influential civic, fraternal, business, and environmental organizations; and formal public workshops and hearings.

Where typical public benefit projects differ from waste-to-energy projects generally centers on the nature and complexity of issues which concern citizen groups. Historically, such projects have attracted significant public opposition due to typical concerns such as: public health and safety; traffic congestion; noise; litter; aesthetics; and property values, among others. Project sponsors need to address these public concerns early during implementation in order to minimize the dissemination of misinformation about the project to reach the general public and thus slow down the project's public approval process. Consequently, it is essential that information and involvement programs be developed early to help elicit public input regarding project decisions, and to respond to public concerns in a timely, but thorough manner. These approaches have historically paid dividends to many projects, which have been successfully implemented, particularly during the period when a project siting decision has its greatest opposition. A number of different strategies have been utilized successfully.

Some project sponsors have presented detailed information regarding the project in informal and formal public forums. Specific citizen concerns expressed at these meetings are noted by project staff.

Where additional technical expertise is needed in the project team to effectively respond to these issues, nationally recognized experts are retained. Subsequent to these meetings, a well-researched response to each of the technical issues raised is presented back to these groups. This may result in modifications or compromises in project recommendations. In this way, the project team is able to successfully defend its recommendations, keep the political process moving along, and win public support.

Other project sponsors have utilized the services of an independent citizen review committee or task force, composed of representatives of various community groups. The purpose of this committee is to provide a forum for resolving issues of public concern regarding the project. These groups have worked well when they have been given explicit charters and specific time frames from the project sponsor.

Such committees provide a means of dialogue between the project sponsor and a group representing community interests to discuss technical, economic and social issues. This forum offers project sponsors the opportunity of responding to citizen concerns by making accommodations - a process which would otherwise be missing in the formal permitting process. By beginning this dialogue prior to the start of permitting activities, project sponsors can often have a better chance of avoiding significant implementation delays.

REFERENCES

1. Berry, Patricia V., and Marc J. Rogoff, Teamwork Plus Communications Equal Success. _Waste Age_, November: 105-108 (1986).

2. Nemeth, Diane M., _The Resource Recovery Option in Solid Waste Management: A Review Guide for Public Officials_, Chicago: American Public Works Association (1981). DOE/CS/20156-T1.

3. Rose, David P., Project Structure-Process and Players. In: _Proceedings of the Third Annual Resource Recovery Conference_, Washington D.C. on March 28-30, 1984, Washington: National Resource Recovery Association.

4. Scaramelli, Alfred B., Resource Recovery Success Depends Upon Commitment. _American City and County_ May: 30-34 (1984).

5. Schoenhofer, Robert F., Project Structure-Process and Players. In: _Proceedings of The Third Annual Resource Recovery Conference_, Washington D.C. on March 28-30, 1984, Washington: National Resource Recovery Association.

6. The Keystone Siting Process Group, _The Keystone Siting Process Handbook_, Austin: Texas Department of Health (1984).

7. U.S. Environmental Protection Agency, *Resource Recovery Management Model - Overview*, Washington: U.S. Environmental Protection Agency (1980). SW-768.

8. Zier, Robert E., Managing Risks Part of Success. *Solid Wastes Management* May: 35-39 (1982).

- 3 -
Waste-to-Energy Technology

INTRODUCTION

One of the first questions a community must answer is what technology will be chosen to convert its solid waste into energy. There is no one "best" technology for everyone. Each community must identify and evaluate the various waste-to-energy technologies that are available and make its own selection based upon the requirements specific to its particular project. This includes consideration of factors (which will be discussed later) such as: available energy and materials markets; the size of the community's waste flow; site availability and location; capital and operating costs; ownership and financing considerations; and the level of risk to be assumed by the community or the facility operator.

In evaluating whether or not one technology better suits its needs than another, a community may often discover that one or more of their goals established for the project may conflict with others. A particular technology, for example, may produce the greatest amount of energy for the community's waste, albeit at the highest projected capital and operating costs. The selection of a technology, therefore, is not a simple one, but one which can require tradeoffs between one goal of a community with others. Since the risks associated with waste-to-energy technology can be substantial, it is critical that each community attempt to minimize these risks as best it can. The following criteria can be utilized to assess the relative riskiness of a particular waste-to-energy technology:

- o <u>Degree and Scale of Operating Experience</u>: Some technologies only have been proven in pilot or laboratory operations, or with raw materials other than municipal solid waste. Other technologies have only been commercially operated in small facilities and the scale up to larger sized plants may result in unforeseen problems;

- o <u>Reliability to Dispose of Municipal Solid Waste</u>: The technology selected must be capable to dispose of solid waste in a reliable manner without frequent mechanical downtimes resulting in diversion of such waste to landfills;

- o <u>Energy and Material Market Compatibility</u>: The technology must be capable of recovering energy and materials for which markets are available;

o <u>Environmental Acceptance</u>: The technology must meet all permitted environmental requirements established by regulatory agencies; and

o <u>Cost to the Community</u>: The technology must dispose of the community's solid waste at a price the community can afford given alternative means of disposal.

This section will briefly describe the basic categories of waste-to-energy technology.

MASS BURNING

"Mass-burning" refers to the generic name for the type of technology used to incinerate unprocessed solid waste, and thereby releasing its heat energy. The thermal reduction of solid waste through mass-burning has been a common procedure throughout the world. There are decades of experience in constructing and operating some 500 mass burn facilities in the United States and Europe. Such facilities were in operation as early as 1896 in Hamburg, Germany, converting solid waste into electricity.

During the period from about 1905 through 1945, there were many overall improvements to these mass incineration systems. The traditional incinerator constructed at this time had boilers which were refractory-lined to protect the outer shells of the boiler from sudden changes in temperature. Furthermore, excess combustion air in the range of 100 to 200 percent above combustion requirements was produced in such units to further cool the walls of the boiler. Unfortunately, the large quantity of excess air produced by these facilities affected the level of pollutants emitted by the refractory-lined incinerators. By the time the Clean Air Act was passed in 1970, this type of incinerator had fallen into disfavor, and many existing facilities were abandoned rather than retrofitted to meet the more stringent air pollution control requirements.

Refractory-lined incinerators also had the additional problem of being less efficient in recovering significant quantities of energy from solid waste. The energy recovered in the combustion process of such incinerators was centered in a waste-heat boiler located downstream from the combustion chamber. The large volume of exhaust gases from combustion pass through this boiler where the heat is absorbed and turned into steam. In such cases, heat recovery is much less efficient than if the boiler were located closer to the combustion chamber.

In the post World War II era, some of the European stoker manufacturers began to experiment with replacing the refractory material in the boilers with waterwall tubing for greater heat transfer which could allow steam to be produced at greater temperatures and pressures. In such incinerators, the walls of the furnace are lined with tubes filled with water, thus the name waterwall incinerators. The world's first integrated waterwall incinerator began operation at Bern, Switzerland in 1954 and continues to operate today. The impetus for this interest in energy recovery from solid waste was the rising cost of energy throughout the industrialized world.

Process Description

An illustration of a typical mass-fired, waste-to-energy facility is shown in Figure 3-1. Solid waste collection and transfer vehicles proceed into a tipping area where their waste is discharged into a large storage pit, which is usually sized to allow two to three days storage or stockpiling of refuse so that plant operations can continue over weekends and holidays when deliveries will not be accepted. There are some facilities which differ in design by utilizing a tipping floor with a front-end loader and belt conveyor system as their form of storage and feed system. In almost all facilities, however, the refuse is fed into the furnaces by means of overhead cranes manipulated by a crane operator. Much of the success of the operation depends upon the skill of the crane operator to remove large or unusual objects in the waste stream that would otherwise prove to be a problem if fed into the boiler. The operator is also responsible to observe the nature of the incoming waste so that materials with different moisture contents are gradually intermixed to try to get a uniform moisture content.

The refuse is then discharged into refuse feed hoppers, which meter out the refuse into the combustion chamber, either by gravity feeding or by a hydraulic feeding device. In a majority of systems, the waste is then pushed onto an inclined, step-like, mechanical grate system which continuously rocks, tumbles, and agitates the refuse bed by forcing burning refuse underneath newly fed refuse. Generally,

36 How to Implement Waste-to-Energy Projects

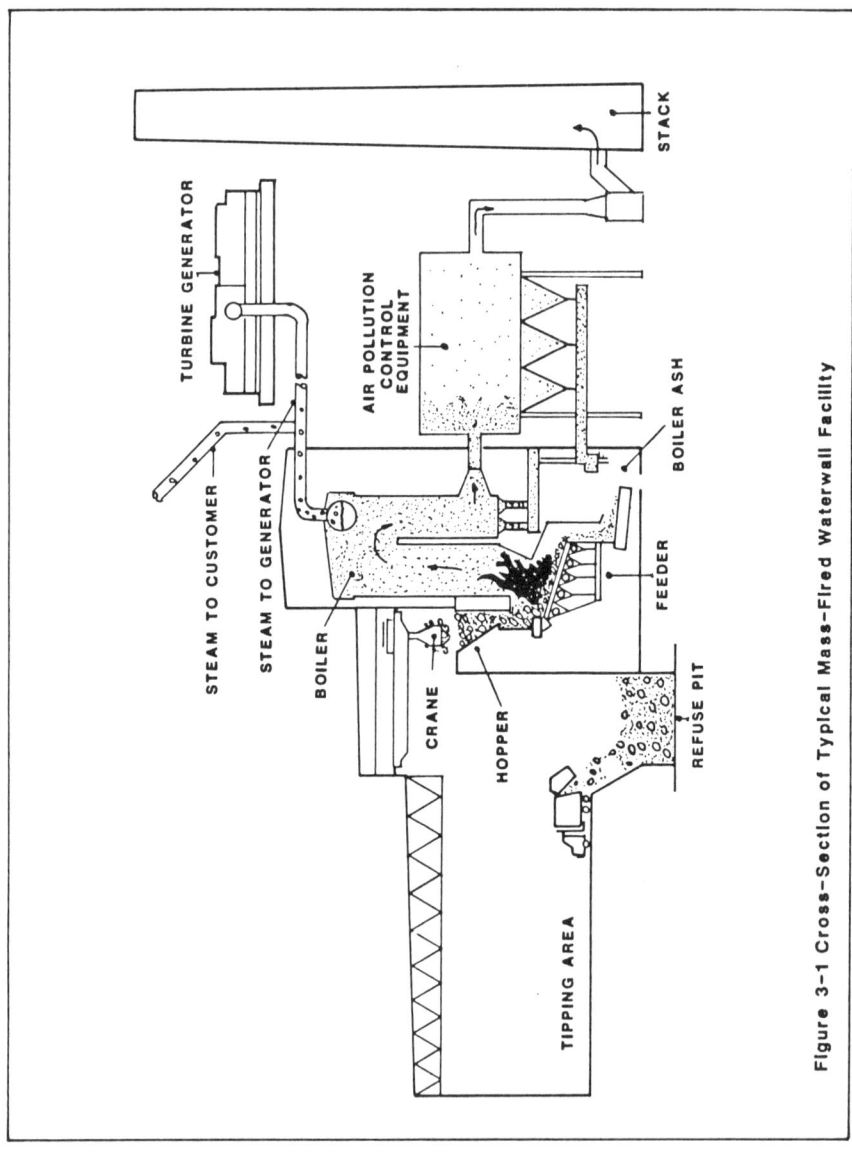

Figure 3-1 Cross-Section of Typical Mass-Fired Waterwall Facility

most systems have three zones of activity along the grates: drying, ignition, and burnout. Holes in each grate bar allow underfire air to pass through the grates resulting in cooling and, thus, preventing thermal damage to the grate system. The width of the grate and the number of grate steps is dependent not only upon the manufacturer's specifications, but also on the overall size of the waste-to-energy system. There are five basic moving grate designs:

- o <u>The reciprocating grate</u>: This grate resembles stairs with alternating fixed or moving grate sections. The pushing action may be in the direction of waste flow or in an upward motion against the waste flow;

- o <u>The rocking grate system</u>: Pivoted or rocked grate sections produce an upward or forward motion, advancing the waste down the grate;

- o <u>The roller grate</u>: A series of rotating stepped drums or rollers agitate the waste and move it down the grate;

- o <u>The circular grate</u>: A rotating annular hearth or cone agitates the waste; and

- o <u>The rotary kiln</u>: As an inclined cylinder rotates, it causes a tumbling action to expose unburned material and advance the waste down the length of the kiln.

Facilities using mass-burn technology

have been designed with either refractory or waterwall furnace systems. The major difference between these systems is the location of the boiler. Refractory units have their boiler located downstream of the combustion chamber, whereas waterwall units have their furnace/boiler units constructed with water tube membrane walls to recover the heat energy. A majority of mass burn facilities constructed have waterwall systems because of their greater thermal efficiency which is generally between 60 to 75%.

In modern waterwall incinerators, proper combustion of the waste is achieved through the introduction of air at two locations in the furnace. One location introduces air underneath the grate system (underfire air) so as to cause agitation and turbulence within the burning waste, and also to help cool the grates. Air is also introduced above the burning waste (overfire air) to ensure that there is adequate oxygen available to completely oxidize and burn all the combustible materials, as well as promote proper mixing of combustion gases. During the combustion process, flue gases, which are heated to temperatures as high as 1,800 degrees F, move from the furnace through the boiler tube sections, such as the superheater section, where the contained water is heated to form saturated steam and dry steam. The flue gases continue through the economizer section to the air pollution control device, such as an electrostatic precipitator, baghouse, or acid gas scrubber, where the flue gases are cleaned before being released into the atmosphere through a stack.

After the combustion process is completed, the grate system or rotary combustor gradually moves the waste onto the burnout grate where it is discharged into a wet or dry ash handling system that cools the residue and prevents dust from being created. The bottom ash that is produced from the combustion process in the furnace, and the fly ash or other materials produced in the air pollution control device, are transported to landfills by truck or to a temporary onsite ash storage pit for later transport. The bottom and fly ash may be combined or handled separately.

Mass burn incineration produces ash residues amounting to 15 to 30% by weight and 5 to 10% by volume of the incoming municipal solid waste. Most facilities can produce an ash product that has less than 5% combustible material and 0.2% putrescible matter.

Recovery of ferrous and non-ferrous materials from the ash residue is possible in mass-burn systems. Many facilities have successfully utilized magnetic separators (with or without trommels) to recover ferrous material from the ash. Some systems have attempted to recover the remaining non-magnetic fraction in the ash, such as aluminum and glass, using various trommels, screens, jigs and fluid separators.

Operational Experience

Mass burning incinerators have been used in Europe and Japan for municipal solid waste disposal for nearly 30 years where their acceptance has been rapid and widespread. With over 500 facilities in operation worldwide in sizes ranging from 60 to 3000 tons per day, mass fired incineration is the most thoroughly demonstrated technology in the waste-to-energy field at this time.

This technology was introduced into the United States in 1967 at the U.S. Naval Station in Norfolk, Virginia with the construction of a 360 ton per day waterwall plant to produce process energy for the naval shipyard. This plant was designed in America and equipped with American equipment. Later plants, which were constructed, were almost entirely designed using state-of-the-art European mass incineration technology. Table 3-1 lists waste-to-energy facilities using mass incineration technology which are currently in operation in the United States at the time of this writing. There are numerous additional facilities currently under construction. The National Resource Recovery Association publishes a semi-annual update of waste-to-energy activities in the United States.

The introduction of European technology into the United States has not been without difficulties and several of the earlier constructed plants encountered some mechanical problems. These highly reliable and rugged European systems had been designed to burn solid waste that was somewhat different in composition than American wastes. Consequently, systems that had been designed for European conditions required designers to make adjustments in the grate areas and furnace heat release rates of American plants. In addition, the higher chloride corrosion of the superheaters in American plants meant that designers needed to change the metallurgy of these boiler tubes, as well as limiting the upper steam pressures and temperatures to minimize tube corrosion. Scale-up problems also had to be overcome

TABLE 3-1
SUMMARY OF OPERATIONAL U.S. MASS BURNING FACILITIES, 1986

Project Location	Year Startup	Units	Design Capacity	Throughput (tons/day)	Capital Cost ($ millions)	Cost Per Ton
Norfolk, VA	1967	2	360	180	2.2	6,111
Harrisburg, PA	1972	2	720	520	8.3	11,527
Chicago, IL	1972	4	1600	1250	23.0	14,375
Nashville, TN	1974	2	720	475	24.5	34,027
Saugus, MA	1975	2	1500	1200	50.0	33,333
Gallatin, TN	1981	2	200	200	10.0	50,000
Hampton, VA	1980	2	200	200	10.4	52,000
Glen Cove, NY	1982	2	250	250	34.0	136,000
Pinellas Cty, FL	1983,86	3	3150	3000	207.0	65,714
Westchester Cty, NY	1984	3	2250	1890	179.0	79,555
Baltimore, MD	1985	3	2250	2250	185.0	82,222
Tampa, FL	1985	4	1000	850	76.0	76,000
North Andover, MA	1985	2	1500	1500	121.0	81,020
New Hanover Cty, NC	1986	2	200	200	14.0	70,000
Tulsa, OK	1986	2	750	500	55.4	73,900
Marion County, OR	1986	2	550	550	47.5	86,363

Source: Reference 10

since many of the European units were designed for the 300 to 500 ton per day range. These problems have been corrected, and most mass-burn systems that have been constructed are still in operation today.

MODULAR COMBUSTION

A modular incinerator is a type of mass-burning, waste-to-energy unit which is prefabricated on a standardized modular basis in a factory. These plants operate on a starved air basis. Such units are shipped to the site in modules, ranging in design capacity from ten tons per day to 200 tons per day, where they are installed. Several modules can be grouped together at a single location. These "off the shelf" units can often be less costly to fabricate than the larger mass-burn facilities which require more costlier field erection. Modular plants can also typically be constructed in some 15 to 20 months.

Process System

Modular incinerators have been designed and constructed in the United States with different process configurations. Some units have been designed to incinerate solid waste under excess air conditions with either refractory furnaces and waste heat boilers or with waterwall boilers. A majority of most units, however, have been designed to operate under starved air conditions with refractory furnaces and waste heat boilers.

A cross sectional view of a typical modular combustion unit is illustrated in

Figure 3-2. A majority of modular facilities have a tipping floor and utilize a front-end loader for simplicity in waste storage and feeding. Combustion takes place in either two or three stages. First, solid waste, which is delivered to the facility, is fed into the initial combustion chamber using a ram-type feeder. A moving ram slides back and forth over fixed steps within the chamber, causing the waste to tumble down one fixed section of the grate to the next fixed section. The waste is then transformed into a low-Btu gas which is then combusted in the secondary chamber, where auxiliary fuel is often fired under excess air conditions. A discharge ram on the back end of the combustion chamber feeds this incinerated waste into an ash quench bath.

The low-Btu gases produced by the combustion process in the first chamber are typically introduced into a secondary chamber where they are burned at temperatures ranging from 1,800 to 2,000 degrees F. Heat energy is recovered by convection in waste heat boilers in this secondary chamber, although waterwall boiler units for the primary and secondary chambers have been constructed.

Operating Facilities

There have been many more modular waste-to-energy incinerators constructed in the United States than either the mass-burn or refuse-derived fuel systems. In 1977, the first modular incinerator began operations in North Little Rock, Arkansas to produce steam for the Koppers Industry's Forest Products Division. Since that time some 50 modular systems have been built in the United States, (Table 3-2), almost exclusively to produce

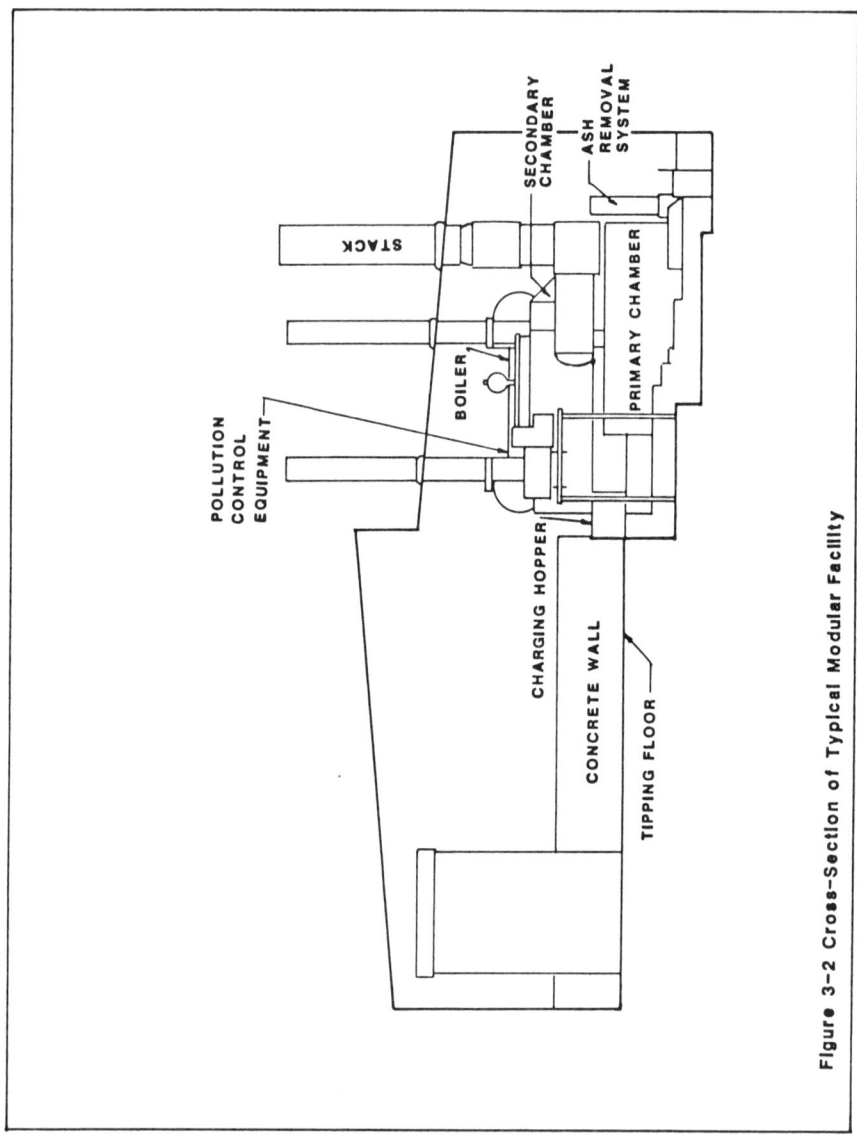

Figure 3-2 Cross-Section of Typical Modular Facility

TABLE 3-2
COMPARISON OF TYPICAL U.S. MODULAR COMBUSTION FACILITIES

Location	Startup	Design Capacity (tons/day)	Energy Generation	Capital Cost ($ millions)	Cost[a] Per Ton
North Little Rock, AR	1977	100	Steam	1.5	14,500
Salem, VA	1979	100	Steam	1.9	19,000
Auburn, ME	1981	200	Steam	4.0	20,000
Durham, NH	1980	60	Steam	3.3	30,555
Osceola, AR	1980	50	Steam	1.2	24,000
Pittsfield, MA	1981	360	Steam	10.8	45,000
Collegeville, MN	1981	64	Steam	2.4	37,500
Windham, CT	1981	108	Electricity	4.1	37,900
Portsmouth, NH	1982	200	Steam	6.3	31,500
Red Wing, MN	1982	72	Steam	2.5	34,700
Cuba, NY	1983	112	Steam	5.5	49,100
Pascagula, MS	1984	150	Steam	5.9	39,300
Tuscaloosa, AL	1984	300	Steam	8.5	28,300
Oswego, NY	1985	200	Steam	14.5	72,500

Source: Reference 10

[a] Estimated cost per ton of design capacity

process team for neighboring industries. Some of these systems, for example, the plant in Collegeville, Minnesota, have utilized the community's solid waste as a fuel to produce steam to a district heating loop during the winter, and electricity during the summer. However, only a few facilities currently generate electricity.

Modular combustion units offer a lower capital cost and simplicity to communities considering waste-to-energy systems than the larger field-erected mass burning systems. These systems are generally reliable and are backed by many years of successful operating experience. Unfortunately, these systems require frequent overhauls of their operating components, which increase maintenance expenses, and result in shorter economic lives for these facilities. Modular units are also less thermally efficient than larger mass-burn units in recovering energy, and thus, are normally used for steam generation only. Another disadvantage is that most units because of their starved air design and less efficient burnout produce more ash ranging from 8 to 15% by weight than the larger mass burn units.

REFUSE DERIVED FUEL (RDF) SYSTEMS

Several American corporations have developed technologies that pre-process solid waste to varying degrees to separate the non-combustibles from the waste stream. By undergoing processing steps of hammering, shredding, or hydropulping, the combustible fraction of the waste is transformed into a fuel, which can then be fired in a boiler unit specifically dedicated for this type of refuse-derived fuel, or co-fired with another fuel, such as coal, shredded tires, or wood chips. The fuel produced can thus be utilized in equipment that can have higher efficiencies than mass-fired units resulting in greater electricity or steam output. However, the front end processing of the solid waste into a fuel has been one of the problem areas of this type of refuse disposal technology.

Processing Systems

The processing of solid waste into a refuse-derived fuel has been approached using both wet and dry processing systems.

Wet RDF Processing. Wet RDF processing utilizes hydropulping technology adapted from the pulp and paper industry. The solid waste is fed into a large pulper, which acts very much like a kitchen blender, where it is mixed with water forming a slurry. The resulting slurry is transferred to liquid cyclones which separate the combustible from the non-combustible fractions. The combustible fraction is then mechanically dewatered and dried to a moisture content of 50 percent solids before being introduced as fuel into a dedicated boiler. The energy efficiency of this process is, however, reduced by the need for drying.

This wet processing system was first utilized in a 150 ton per day pilot plant in Franklin, Ohio from 1972 to 1979, which is now closed. The first full-scale 2,000 ton per day system was constructed by Black Clawson, a subsidiary of Parsons and Whittemore Company, at Hempstead, New York. This plant experienced several early operational difficulties resulting in its closure. There are plans to demolish this facility and construct a mass-fired, waste-to-energy energy system on the existing site. A 3,000 ton per day sister facility, is currently in operation in Dade County, Florida. The operators of this facility, however, have reportedly considered abandoning the hydropulping equipment.

Dry Processing Systems. Since the early 1970's, there have been several dozen facilities which have been constructed in the United States to process solid waste into a refuse-derived fuel through the use of dry processing systems. Such dry processing systems are classified according to the type of products that can be produced: fluff RDF, densified RDF, and powdered RDF. A cross-section of a typical RDF system is illustrated in Figure 3-3.

Given the number of potential products, the type of specific technology used to process the solid waste into a refuse derived fuel can vary from one location to another. Typically, however, solid waste delivered to an RDF facility is unloaded onto the receiving floor, or at some locations into a refuse storage pit. The waste is then transported with the use of a feed conveyor system to a size reduction unit, which reduces the particle size of the waste. At many facilities, machinery such as flail mills, trommels, and magnetic separators, are used to pre-sort the waste prior to its being fed into a hammermill or shredder for size reduction. Depending on the particular type of RDF fuel required, further processing equipment is utilized after shredding, such as air classifiers,

Waste-to-Energy Technology 49

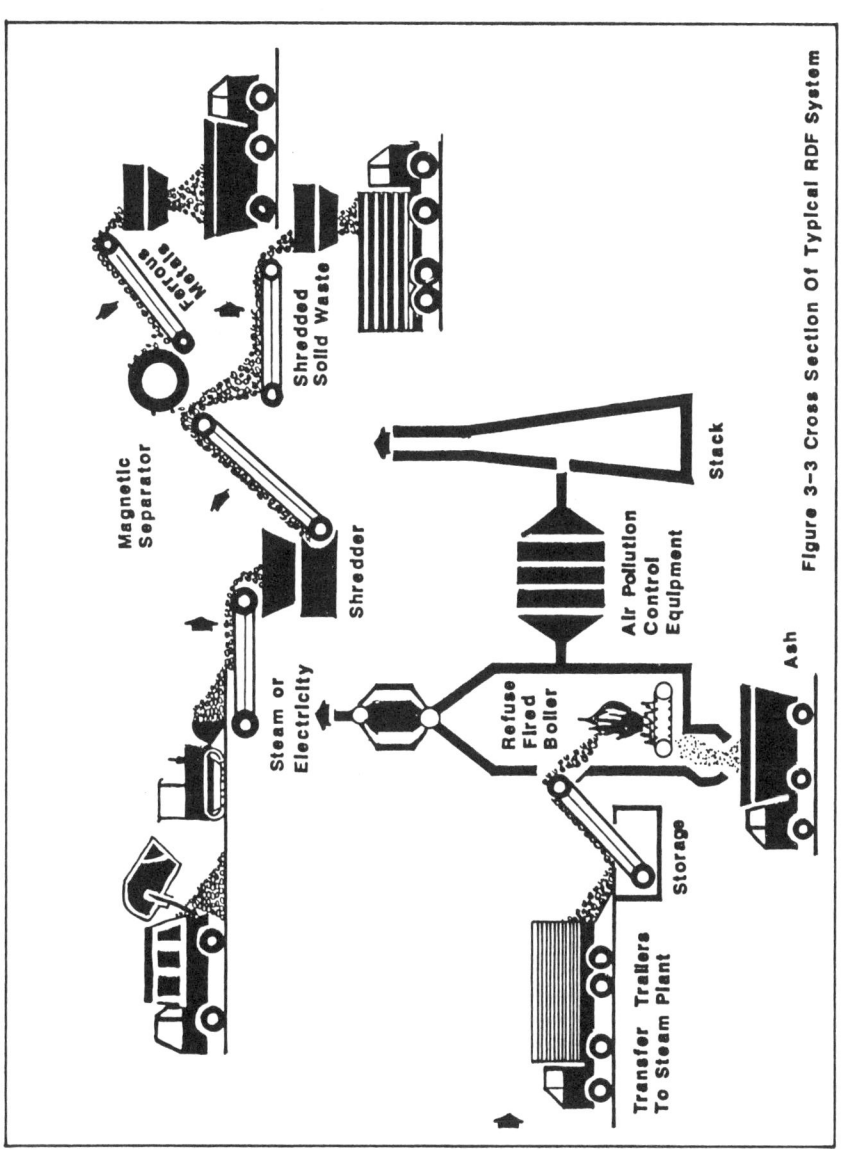

Figure 3-3 Cross Section Of Typical RDF System

densifiers, and trommel screens. The end result of the dry processing system is a refuse-derived fuel which can be combusted in either existing utility boilers, or in boilers specifically designed for the type of RDF produced (dedicated boilers).

For many years, the plan of burning RDF in existing electric utility boilers seemed an obvious solution for communities which needed a good way to dispose of garbage. It was hoped that existing boilers and air pollution control equipment could be utilized thus saving these communities considerable capital expense. Since the early 1970's, RDF has been tried, however, as a supplemental utility fuel with mixed success. Since 1970, eight utilities in the United States have co-fired RDF in their system boilers, only three are still burning RDF.

Shortly after the beginning of the first demonstration project in St. Louis in 1972, it became apparent that burning RDF in utility boilers resulted in a lowering of their normal efficiency and reliability. When RDF was fired in the high temperature, utility boilers, the non-combustible materials in solid waste, such as glass and metals, melted into slag that fouled the boiler tubes, heat exchangers and furnace walls. Burning of the plastic compounds in the solid waste, which released chlorine, also resulted in increased corrosion of boiler parts. In addition to these problems, ash handling, air pollution control, and materials handling systems soon became overloaded and were subject to frequent outages. In short, what had seemed to be a good way to dispose of solid waste resulted in an unexpected headache for utilities. The initial optimism of this technological fix to solve an emerging garbage disposal problem has consequently not been realized.

The recent emphasis on burning RDF has focused on systems using "dedicated" industrial type boilers. The term, "dedicated", refers to a boiler system that is specifically designed and constructed to burn RDF as its primary, not supplemental, fuel. There are a variety of different types of technologies which had been used in such boilers: suspension-fired; semi-suspension fired (spreader stoker); pyrolysis; and fluidized bed.

The semi-suspension fired or spreader-stoker, furnace boiler is perhaps today's most commonly utilized technology. Spreader-stoker technology has been utilized successfully for decades for incineration of a variety of different solid fuels. With this system, RDF, which has been previously prepared to specific size characteristics, is introduced at a controlled rate to pneumatic RDF distributors located at the front wall of the furnace. High-pressure air is delivered to these distributors to assure that the RDF is fed evenly. The RDF, so introduced, ignites over the grate area and burns partially in suspension. Materials, which are left unburned, fall to the traveling stoker where they are combusted before the ash is discharged.

Table 3-3 lists RDF fired, spreader-stoker systems, which are currently in operation, and are dedicated exclusively to firing RDF. There are several projects under construction which will utilize spreader-stoker boilers. The experience with these RDF systems has varied. It

TABLE 3-3
SUMMARY OF OPERATIONAL U.S. RDF FACILITIES, 1986

Location	Startup	Design Capacity (tons per day)	Capital Cost ($ millions)	Type of Process
Ames, IA	1975	200	6.3	Shredded, air classified; ferrous recovery; RDF co-fired with coal
Akron, OH	1979	1000	80.0	Shredded, air classified; ferrous recovery
Albany, NY	1981	750	28.2	Shredded, air classified; ferrous, non-ferrous recovery from boiler ash; RDF preparation plant and boiler plant on separate sites
Cockeysville, MD	1976	1200	11.0	Shredded, air classified; ferrous recovery; co-fired with coal
Columbus, OH	1983	2000	175.0	Shredded; ferrous recovery
Duluth, MN	1981, 85	400	19.0	Shredded, disc-screened; co-disposal with sewage sludge; reopened after changes to front-end processing system; fluidized bed boiler
Dade Cty, FL	1982	3000	165.0	Wet processing; ferrous, glass, nonferrous recovery

TABLE 3-3 (Continued)

Location	Startup	Design Capacity (tons per day)	Capital Cost ($ millions)	Type of Process
Haverhill, MA	1984	1300	99.5	Shredded, air classified; ferrous recovery; RDF preparation and boiler plants on separate sites
Lakeland, FL	1983	300	5.0	Shredded; air classified; ferrous recovery; co-fired with coal
Madison, WI	1976	400	2.5	Shredded; air classified; ferrous recovery; co-fired with coal
Niagara Falls, NY	1980	2000	100.0	Shredded; air classified; ferrous recovery
Wilmington, DE	1984	1000	72.3	Shredded; air classified; ferrous non-ferrous, glass recovery; humus (co-composting); no combustion

Sources: References 8 and 10

appears that most installations have had problems with RDF feeder equipment resulting in extensive retrofits and technical modifications. However, where high quality RDF has been introduced with most metals and glass removed, the RDF burning experience has generally been good. On the other hand, the experience of co-firing of RDF with coal has been generally poor. The combination of coal and RDF appears to increase problems with ash clinker and slagging and wear of the lower furnace walls.

FLUIDIZED BED SYSTEMS

Fluidized-bed technology has been in existence for well over 50 years, principally in Europe. This technology has been utilized for burning a variety of low-quality fuels such as: high-sulfur coal; peat; tires; sludges; waste oils; and biomass wastes. Within the last decade, fluidized-bed technology has undergone extensive refinement.

Three types of systems have been developed for burning coal: the bubbling-bed; the dual-bed; and the circulating-bed. For combusting solid waste, the circulating-bed process has proven to be the best method. Utilizing this system, refuse derived fuel would be injected into a fluidized-bed combustor which is composed of a steel and refractory brick liner and a fluidized bed of limestone and sand. The heated bed of thermally inert materials behaves like a fluid because high-velocity air is injected into the combustion chamber. The RDF is introduced into or on top of this circulating-bed where it is combusted at temperatures typically in the range of 1,550-1,650 degrees F. Energy from the flue gases is removed using either a waste heat boiler, heat exchanger tubes, in-bed tubes, or waterwalls.

Although fluidized-bed technology has been utilized in the utility industry, there has been little demonstrated experience in the United States to date with municipal solid waste. European experience in fluidized-bed incineration of solid waste consists of several plants, principally in Sweden, where RDF is incinerated in conjunction with other fuels. Limited operational data are available because these facilities have only begun commercial operation within the last few years.

In the United States, fluidized-bed incineration of RDF has been tested at a number of pilot locations in Menlo Park, California; at West Virginia University; in Franklin, Ohio; and at French Island, Wisconsin. A commercial-scale system for co-firing refuse and sludge was constructed in Duluth, Minnesota. A fluidized-bed system has been proposed in Collier County, Florida and Erie, Pennsylvania.

The experiences at these pilot projects and in Duluth indicate that the use of fluidized-beds to incinerate municipal solid waste is in a research and development phase. Success at these projects has been shown to depend on maintaining a stable bed of limestone, sand, and RDF. Unfortunately, experience has shown that the ash and residue from the RDF fuel can change the physical and chemical composition of this bed, necessitating that the bed be continuously replaced. A further problem that has been encountered is non-combustible particles in the RDF, such as glass and metals, can accumulate in the bed and fuse with the ash. The result is that the bed becomes de-fluidized. To overcome these problems, extensive processing of the RDF fuel is necessary. The degree of the processing required could negatively impact the operational economics of such systems. However, such systems have shown that acid gases can be removed with considerable efficiency. Thus, expenditures for air pollution control equipment could be expected to be significantly reduced.

ANAEROBIC DIGESTION

Under this process, organic materials in a sanitary landfill are decomposed in the absence of oxygen. A demonstration project in Pompano Beach, Florida, funded by the U.S. Department of Energy, has been underway for several years to test the theory that methane gas can be produced under controlled anaerobic conditions in a landfill composed of solid waste mixed with sewage sludge. This technology currently has not been demonstrated for commercial application.

COMPOSTING

Composting involves the aerobic decomposition of organic matter by microorganisms to form a stable humus-like material which can be used as a soil conditioner. The practice of placing organic wastes such as animal manures, kitchen wastes, yard wastes, and leaves in large piles or pits for aerobic decomposition has probably been practiced for centuries. In addition, the composting of sewage sludge and selected organic materials such as leaves, tree trimmings, and wood chips has been carried out successfully for many years in the United States.

Commercial production of compost in the United States has been undertaken in nearly 200 mechanical composting plants. Unfortunately, much of the experience of such facilities is with sewage sludge rather than with municipal solid waste. Composting municipal solid waste is considerably more difficult because it contains materials which are unable to be composted, such as: metals, glass, and plastics. These materials must be separated before composting with a processing system similar to that required to produce RDF. There are a number of different systems being promoted to process municipal solid waste into compost. Each system differs in the way it separates the non-organics in solid waste from the organic materials and then accelerates decomposition of these organic materials.

Over the last 30 years, some 24 facilities for composting municipal solid waste were constructed in the United States. Most were operated for only a short time, primarily because the resulting compost product had poor marketability. Some of the other reasons for plant closings included: limited solid waste disposal capability; odor problems; heavy fly infestation; and contaminants in the product. Currently, the only operating plants in the United States are located in Albany, New York and Wilmington, Delaware. The composting operations at these two RDF facilities are relatively small in volume. The Albany plant only composts when there is no market for its RDF. The Delaware plant composts sewage sludge and the fines from the RDF facility.

In summary, composting of municipal solid waste with sewage sludge appears to have some potential as an alternative means of solid waste disposal. Unfortunately, the track record for this emerging technology to date has been poor.

PYROLYSIS CONVERSION SYSTEMS

The pyrolysis of solid waste to produce liquid or gaseous fuel received much attention in the United States following the energy crisis in the 1970's. During this time, several companies such as Union Carbide, Andco-Thorax, Occidental and Monsanto developed different types of pyrolysis systems. Demonstration systems were built in Baltimore, Maryland; San Diego, California; and Charleston, West Virginia. These facilities were unable to demonstrate economic feasibility on a full-scale basis. This led to the demise of pyrolysis as a commercially available technology receiving serious attention.

COMPARISON OF TECHNOLOGIES

The two most prevalently used technologies for waste-to-energy facilities in the United States--mass burning and prepared fuel systems-- can be compared with respect to efficiency in energy recovery and cost.

Energy Efficiency

Table 3-4 provides a comparison

TABLE 3-4
COMPARISON OF TYPICAL ENERGY RECOVERY EFFICIENCIES

Technology	Fuel Heating Value (BTU/lb)	Boiler Efficiency (%)	Steam Produced (lb/ton)	Gross Electrical Output (kwh/ton)	Plant Power Consumption (kwh/ton)	Net Electrical Output (kwh/ton)
Mass Burn						
Field Erected	4500	60-70	4500-6000	470-580	55-70	400-520
Modular	4500	55-60	4000-5000	350-440	45-55	300-375
RDF[a]	6000	74-78	5400-5800	470-570	80-90	450-500

[a] Dry RDF in a dedicated boiler.

between these two classes of waste-to-energy technologies. As indicated, prepared fuel systems with dedicated boilers have higher boiler efficiencies and produce significantly larger quantities of steam than mass burn systems. However, RDF systems consume more energy than mass burn systems, primarily for powering the equipment for processing solid waste into a prepared fuel. Similarly, field erected, mass burn facilities have been shown to exhibit higher boiler efficiencies and produce more quantities of steam and net electricity than modular facilities.

Cost

Waste-to-energy facilities vary with respect to their capital and operating costs. Each project can be considered unique. The range in project costs are due primarily to a great number of design and site specific issues such as: equipment requirements; construction details; local labor rates; taxes; energy market needs; and particularly the methods of procurement, plant ownership and financing.

Historically, there are economies of scale for a community to construct and operate one large facility as opposed to several smaller, and separate facilities, for example, one 1,000 ton per day facility instead of two 500 ton per day plants. This is a reasonable conclusion since the community would probably have to duplicate similar equipment, facilities, and manpower at each plant, which would inflate overall project costs. Yet, it would also be expected that these economies of scale would diminish dramatically once this single community plant increased over a certain size. This would be due in part to problems associated with moving vehicles and materials to and from the facility.

The capital costs associated with constructing modular facilities are also historically less than similar field erected, mass burn plants. This cost differential has resulted from the following reasons:

- o Standardized factory parts simplify field erection, thereby reducing costs;

- o Structural requirements for supporting smaller pieces of equipment are significantly less;

- o Permitting requirements have been less stringent for these sized facilities;

- o The starved-air design reduces the size of post-combustion air pollution control equipment;

- o The smaller volumes of waste require simpler materials handling equipment;

REFERENCES

1. Battelle Columbus Laboratories, <u>Refuse Fired Energy Systems in Europe: An Evaluation of Design Practices - An Executive Summary</u>, Washington: U.S. Environmental Protection Agency (1979). SW 771.

2. Engdahl, Richard D., Energy Recovery From Raw Refuse Versus Refined Municipal Wastes. In: **Proceedings of the 1986 National Waste Processing Conference**, New York: American Society of Mechanical Engineers (1986), pp. 105-112.

3. Goldstein, Nora, Sewage Sludge Composting Maintains Momentum. **Recycling** November/December: 21-26 (1986).

4. Miliaras, E. Stephen and David Child, Optimizing the Use of Refuse Derived Energy in Electric Utility Systems. Speech presented at Energy From Municipal Wastes: Opportunities in an Emerging Market in Washington, D.C. on October 24-25, 1985.

5. Niessen, Walter R. and Thomas C. Pond, Modular Combustion Units. **Public Works Magazine** May: (1980).

6. Peterson, Charles and Raymond Givonetti, Municipal Solid Waste for Energy: A Technology Review. Speech presented at 11th Energy Technology Conference in Washington, D.C. on March 21, 1984.

7. Smith, M.L. One View of RDF Options. **Waste Age** April: 120-130 (1986).

8. State of Florida, **Modular Incinerators**, Tallahassee: Department of Environmental Regulation (1983).

9. Systech Corporation, *An Overview of U.S. and Canadian Experience With European Mass Burning Waterwall Incineration Systems*, Washington: U.S. Environmental Protection Agency (1981).

10. U.S. Conference of Mayors, Report on Semi-Annual Survey: Resource Recovery Activities. *City Currents*, April (1986).

11. U.S. Environmental Protection Agency, *A Guide to Energy From Municipal Waste For Small Communities*, Washington: U.S. Environmental Protection Agency (1979). SW-177c.

12. U.S. Environmental Protection Agency, *Resource Recovery Plant Implementation - Guides for Municipal Officials: Technologies*, Washington: U.S. Environmental Protection Agency (1976). SW-157.2.

13. U.S. Environmental Protection Agency, *Small Modular Incinerator Systems With Heat Recovery - A Technical, Environmental and Economic Evaluation*, Washington: U.S. Environmental Protection Agency (1979). SW-177c.

- 4 -
Solid Waste Composition and Quantities

INTRODUCTION

Proper planning of a waste-to-energy facility requires that a reliable data base be available on solid waste characteristics and quantities currently generated and expected to be generated within the service area of the facility. Such data is necessary, not only for determining the current refuse disposal needs of the community, but also to determine the overall future requirements of the solid waste disposal system. The quantities of solid waste generated by the community may impact the initial sizing of proposed waste-to-energy facilities, transfer stations, emergency/ash residue landfills and other ancillary facilities, as well as determining the most efficient location of such facilities, so as to minimize transportation costs. The quantities of solid waste generated by the community and available for processing by the waste-to-energy facility is also an important factor in determining the financial feasibility of a proposed facility since the revenues generated through the sale of energy and recovered materials are directly correlated to the amount of solid waste received by such facilities. The composition of a community's waste is also a critical factor since it can affect the energy content of the waste received by a waste-to-energy facility, as well as the quantities of recyclable materials and residues that may be generated. Thus, waste composition can influence the design criteria and economics of any waste-to-energy facility.

CHARACTERIZATION OF SOLID WASTE

For the purposes of our discussion, municipal solid waste refers to solid or semi-solid discarded materials resulting from industrial, commercial, agricultural, institutional, and residential operations, but does not include solids or dissolved materials in industrial or domestic sewage (waste sludges), or hazardous wastes. The latter wastes are now not permitted to be disposed of at solid waste facilities by federal and state law because of their potential impacts on the environment. These materials may include volatile, chemical, biological, explosive, flammable and radioactive materials. While solid waste from most communities contains small amounts of such similar materials as solvents, paints, and household cleaners, they constitute a minimal potential impact upon the environment since they are a relatively small fraction of the total volume of waste generated by most communities. Consequently, federal and state regulatory agencies do not currently consider municipal solid waste as hazardous.

The following classifications of municipal solid waste materials will be used as a basis for subsequent discussion:

- o Residential waste: These are household materials generated by residents in single-family and multi-family dwellings and low, medium and high-rise apartments. These mixed household wastes include kitchen and food wastes (often called garbage) that are highly putrescible and will decompose rapidly posing a health and safety problem if left uncollected. Solid waste generated by households also includes combustible materials such as paper, cardboard, plastics, and garden trimmings, and non-combustible materials such as glass, metals, and soil.

- o Commercial waste: These wastes are generated by wholesale and retail stores, restaurants, markets, office buildings, hotels, motels, and other similar establishments, and large institutional facilities such as hospitals, prisons, schools, and religious institutions. The wastes generated by the commercial sector are very similar in physical characteristics and composition to that generated by residential units.

o Industrial wastes: These are solid wastes generated by various types of manufacturing and industrial operations, excluding such hazardous materials such as solvents, oils, chemicals, and similar manufacturing establishments.

o Special wastes: These are solid wastes which because of their physical characteristics require special or extraordinary handling. This includes bulky materials such as abandoned vehicles, used tires, white goods (e.g., refrigerators, washing machines, etc.), and furniture, and materials generated from demolition and construction projects such as soil, stones, concrete rubble, bricks, lumber, and shingles.

WASTE COMPOSITION STUDIES

Information on the composition of solid waste is essential in evaluating the equipment needs for a waste-to-energy system; establishing the energy yields from the combustion of the waste; and for identifying waste stream components that may be recoverable producing additional revenues for the project. The quantity of ferrous and non-ferrous metals available in the waste stream, for example, may dictate the feasibility of installing a metals recovery system as part of a waste-to-energy project.

As would be expected, solid waste composition varies from community to community due to geographic location, life style, daily habits, season, and climate. While it is impractical to sample the total solid waste stream of any one community, it is possible to develop a sampling program that approximates the composition and variability of a community's waste stream. Through the use of a random sampling technique, a representative sample of the waste stream can be selected for the purposes of the survey.

There are many sampling techniques that may be utilized. Typically, waste samples of approximately 200 to 300 pounds are collected from a specific number of the community's public and private waste collection routes to ensure the statistical significance of the data obtained in the survey. Some sampling programs have utilized a stratified random sampling procedure in order to select the collection routes proportional to the percentage of the community's municipal or private pick-up, or the percentage of residential versus commercial solid waste routes. Once the routes are selected, collection vehicles are diverted to an unloading facility, such as a transfer station, where representative samples of refuse are collected from each of the vehicles. These samples are then separated and hand sorted into various categories, and then weighed to determine a percentage composition.

However representative of a community's waste stream, a waste composition study only determines the composition during the specific sampling period. Since waste-to-energy facilities must be designed to operate year round with seasonal variations, it is important to sample the waste stream at least on a quarterly basis. Many sampling programs have attempted to account for seasonal fluctuation in the composition of the waste stream by sampling the waste stream during one full week each quarter (e.g., November, February, May, August).

Table 4-1 provides some composite results of waste composition studies undertaken by communities in the United States. These are intended only to provide some basic data since the waste stream composition of an individual community may vary significantly. In addition, a community looking at some type of materials recovery may desire to sample for additional categories such as: corrugated paper; newsprint; magazines; aluminum; and other non-ferrous materials.

HEATING VALUE

Since the sale of energy plays an important role in the economic feasibility of a waste-to-energy project, the heating value of the waste stream is a key design factor. The heating value of solid waste is measured by British Thermal Units (Btus) per pound of refuse. One Btu is defined as the amount of energy required to raise the temperature of one pound of water one degree Fahrenheit. Thus, the heating value is a basic measure of the heat energy released through the incineration of solid waste. Typical heating values of municipal solid waste range from 3,000 to 6,000 Btus per pound with 4,500 Btu per pound generally cited as an average.

TABLE 4-1
COMPOSITION OF TYPICAL MUNICIPAL SOLID WASTE STREAM
(Percent As-Disposed Weight)

Material Categories	Ranges	National Average
Combustibles:		79
Paper	64.2 – 94.7	40
Food Wastes	36.6 – 43.9	17
Yard Wastes	13.3 – 20.2	13
Wood	7.9 – 17.4	3
Textiles	2.5 – 3.7	2
Rubber and Leather	1.6 – 2.7	2
Plastics	1.2 – 2.7	2
	1.1 – 4.1	
Non-Combustibles:		21
Metals	18.3 – 24.9	9
Glass	8.4 – 10.1	9
Inorganic Wastes	8.4 – 10.3	3
	1.5 – 4.5	

Sources: References 1, 3 and 4

There are several types of methods which are used to determine the heating value of a community's waste stream. One method is to place a small representative sample of waste in a sealed container, called a bomb calorimeter, in which the solid waste is burned with excess oxygen to ensure complete combustion. The heat of combustion is then measured. Some data indicate that this technique may overestimate the heating value of municipal solid waste.

The other technique is to calculate the heating content of the waste based on fixed carbon, volatile solids, and moisture content. Using this technique, a representative sample of waste is dried to a constant weight in an oven, and, in a further step, reduced to ash in a high-temperature furnace. Weight measurements between steps reveal the moisture content, volatile solids, fixed carbon, and ash of the sample.

SOLID WASTE QUANTITIES

The quantities of solid waste generated by a community can be determined from a number of data sources. The primary source of data are records kept by public and private landfill operators in the community over several years. These data, however, may be incomplete due to poor or variable recordkeeping from one location to another. In addition, it may be impossible to accurately determine the full waste load of certain waste collection routes in a community. Furthermore, the waste disposed of at such locations may be recorded on a per cubic yard or volume basis rather than being weighed on a scale. Additionally, waste may also be disposed of illegally, never entering the community's waste disposal system. These problems are not uncommon throughout the United States, and estimating procedures must be utilized to develop reasonable solid waste generation rates for the community.

Population data are useful to help compare existing records of solid waste tonnages for a community. A per capita solid waste rate can be developed which can then be compared to generation rates for other communities with similar characteristics to determine reasonableness of such a solid waste generation rate. In some instances, it may be useful to develop per capita generation rates for different solid waste classifications (e.g., residential, commercial, and industrial) to assist in this comparison and to determine the reasons, if any, for differences between the community's per capita generation rate and other similar communities. In this way, high or low rates may be easily explained due to certain specific local factors. The key point of such an analysis is to determine a reasonable solid waste generation rate which the community can guarantee to deliver to a waste-to-energy facility, and upon which long-term waste projections can be based. In this way, the waste-to-energy facility can be prudently designed and sized to accommodate future expansion due to population growth.

REFERENCES

1. Brown and Caldwell, <u>Resource Recovery Implementation Planning-</u>

Hillsborough County, Florida, Tampa: Hillsborough County, Florida (1981).

2. Child, David, Gail Ann Pollette and Herbert W. Flosdorf, Waste Stream Analysis. Waste Age, November: 183-192 (1986).

3. National Solid Waste Management Association, Basic Data: Solid Waste Amounts, Composition and Management Systems, Washington: National Solid Waste Management Association (1985). Technical Bulletin 85-6.

4. Smith, Omar E. and Joseph E. Royer, Garbage Coast to Coast: Waste Composition Studies in Three Communities, Memphis: Leonard S. Wegman Co., Inc. (1982).

5. Tchabanoglous, George, Hilary Theisen and Rolf Eliassen, Solid Wastes: Engineering Principles and Management Issues, New York: McGraw Hill Book Company (1977).

- 5 -
Waste Flow Control

INTRODUCTION

One of the more critical issues facing public officials pursuing waste-to-energy facilities is what is commonly termed "waste flow control". In essence, each community must be able to assure those who will be operating its waste-to-energy facility and the financial underwriters for such a project that the solid waste generated from residential, commercial, and industrial establishments within the community will be available on a long-term basis to support a waste-to-energy facility. Without strong control of the solid waste stream, there is the potential for diversion of solid waste from the community's facility. This would be an unacceptable situation because the revenues from tipping fees and the sale of electricity, steam, or recovered materials are used to finance the construction and long-term operation of such facilities.

Waste stream control has been an issue of controversy in recent years between the waste-to-energy industry and

local governments on one hand and the
solid waste haulers and the waste
recycling industry on the other hand.
This latter group has argued against the
imposition of monopolistic waste flow
control by local government for waste-to-
energy facilities since this would
interfere with interstate commerce and
severely restrict their long-term
financial liability by restricting their
continued access to recyclable materials
taken from the waste stream. Spokesmen
for the group have asserted that recycling
of materials from a community's waste
stream would be beneficial rather than
detrimental to the financial integrity of
waste-to-energy facilities because the
size and capital costs of such facilities
could be reduced through initiation of
flow reduction programs.

   This argument has been rejected by
many communities and the investment
community. Spokesmen for these groups
have argued that the financing of waste-
to-energy facilities can not take place
without the long-term assurance on the
part of government that a community's
solid waste is committed for delivery to
the waste-to-energy facility. Without
such assurance, the investment community
has asserted that the interest rate for
project financing would increase
dramatically. Furthermore, some
representatives of local government have
asserted their rights to prohibit
scavaging of materials at the curbside
because of public health and safety
considerations. Some communities in
recent years have attempted to take a
middle course by enacting waste flow
ordinances with commitment for waste-to-
energy facilities, while at the same time
encouraging the development of a strong

recycling industry in their community. Waste reduction and the development of waste-to-energy projects need not be incompatible.

FLOW CONTROL MECHANISMS

This chapter will discuss the three general types of waste flow control mechanisms prevalent in the United States: (1) legislative supplemented by enforcement; (2) contractual; and (3) economic or cost incentives.

Waste Flow Control Through Legislation/Regulation

Local government can exercise some type of legal or regulatory authority over the collection, removal, and disposal of solid waste in its area of jurisdiction. Courts have long upheld the right of governments to adopt reasonable regulations in this area since all property rights were considered to be superceded by local government's police powers. Most of the court cases involving solid waste were decided by jurists at the turn of the century on the premise that regulatory authority was essential to public health and safety, since without such control solid waste would become a nuisance to neighboring property owners.

Perhaps the most important legal decision in recent years regarding solid waste flow control by local government involves the City of Akron, Ohio. Under the terms of the bond convenants for the $46,000,000 bond issue of its 1,000 ton per day, waste-to-energy facility, the City of Akron was required by its bond underwriters to enact an ordinance that would do the following: guarantee a supply of solid waste to the facility by prohibiting the establishment of alternative solid waste disposal sites; required all garbage collectors within Akron and Summit County, Ohio to deposit all waste acceptable for disposal at the plant (including recyclables); and required all collectors to pay a tipping fee when they deposited solid waste at the facility. Violation of this city ordinance by a solid waste collector would result in possible loss of his license and make him subject to criminal penalties.

Prior to enactment of this ordinance, private collectors in the City of Akron and Summit County were able to shop around for solid waste disposal sites with the best disposal price, and also were able to recover and sell valuable recyclables from the solid waste stream before taking the remainder to the landfill. The imposition of waste flow control in the Akron, Ohio area interfered with the previous operations of the private collectors and landfill operators by substantially reducing their incomes.

In <u>Glenwillow Landfill, Inc. v. City of Akron, Ohio,</u> 485 F. Supp. 671 (ND Ohio 1979), these groups argued at the Federal District Court level that the City solid waste control ordinance violated due process; took private property in violation of the Fifth Amendment of the U.S. Constitution; illegally restrained interstate commerce allowed under the Commerce Clause of the U.S. Constitution; and violated the Sherman Anti-Trust Act. In ruling for the City of Akron, the court found that the ordinance was a proper exercise of the city's police powers, and as such did not result in a taking for which compensation must be paid under the Fifth Amendment of the U.S. Constitution.

The District Court's ruling was appealed to the U.S. Court of Appeals, Sixth Circuit. In <u>Hybrid Equipment Corp. v. City of Akron,</u> Ohio, 654 F. 2d 1187 (6th Circuit 1981), the Circuit Court upheld the City of Akron's waste control ordinance. Citing two U.S. Supreme Court cases: <u>California Reduction Company v. Sanitary Reduction Works,</u> 199 U.S. 306, 26 S.Ct. 100, 50 L.Ed 204 (1905) and <u>Gardner v. Michigan</u>, 199 U.S. 325, 26 S. Ct. 106, 50L.Ed 212 (1905), the Court found that the City had not violated the due process and taking clauses of the U.S. Constitution, and the ordinance was a proper exercise of the traditional exercise of local governments police powers. The Court also ruled that the City's actions did not seriously burden interstate commerce, and that the City was exempt from the Sherman Anti-Trust Act. The plaintiffs in this case then appealed to the U.S. Supreme Court.

However, before this case reached the Supreme Court, the Court had handed down a ruling in <u>Community Commmunications Company, Inc. vs. City of Boulder, Colorado,</u> 455, U.S. 40, 102 S. Ct. 835, 70 L. Ed 2d 810 (1982) holding that a municipality can be held responsible for violations of the federal antitrust laws unless it is acting pursuant to "...a clear and affirmatively expressed state policy" permitting such restraint of trade. In light of its decision in the Boulder case, the Supreme Court overturned the judgement against the plaintiffs in the Akron case (solid waste haulers) and sent the case back to the Circuit Court to reconsider the issue of state action exemption for local government under the Sherman Anti-trust Act. The Sixth Circuit remanded this case to the Federal District Court for disposition.

In reviewing the facts of the case, the District Court ruled that the City of Akron was exempt from antitrust liability because the City was acting in furtherance of "clearly articulated and affirmatively expressed" policies of the State of Ohio in the financing of waste disposal facilities. The Court found that the State Legislature of Ohio had contemplated the use of anti-competitive measures to ensure the financial liability of its waste disposal facilities. Consequently, the Court reasoned that as long as local government was acting pursuant to a clearly articulated and affirmatively expressed state policy that indicates an intent of the legislature to displace competition with regulation, local government would be exempt from antitrust liability. Furthermore, the Court, relying on Town of Hallie vs. City of Eau Claire (No. 82-1715, Slip Op. 7th Circuit Court February 17, 1983) held that the active state supervision requirement for antitrust immunity does not apply to municipalities engaged in a traditional municipal function authorized by the state.

Prior to the final disposition of the Akron case, the financial and legal advisors to communities who were hoping to implement waste-to-energy facilities insisted that some sort of legal mechanism be used to confer state immunity from antitrust actions upon local government. To achieve such immunity, however, it was believed that local government needed to meet the "active" state supervision test.

Some local governments have attempted to demonstrate such state oversight of their waste-to-energy programs by having the state, through special legislation, officially delegate the power of supervision to them. Others through special legislation have reaffirmed the existing state supervisory powers over its waste-to-energy systems, including the issuance of permits and periodic reviews.

The Akron decision makes it clear that waste flow control ordinances for waste-to-energy facilities must be carefully drafted by local government. They must balance the needs of government to assure secure waste supplies for its waste-to-energy facility against the legitimate economic concerns of collectors and the recycling industry to remove recyclable materials from the waste stream. Waste-to-energy facilities and recycling need not be incompatible. Many local governments, while enacting strong flow control ordinances, now permit the recovery of recyclable materials in the waste stream.

Contractual Control of Waste Stream

Rather than resorting to the enactment of waste stream control legislation, local government can assure adequate quantities of solid waste for its waste-to-energy facility through contractual controls. This is accomplished when local government enters into long-term contracts with other local governments and private collectors to deliver solid waste to a waste-to-energy facility. This method of voluntary contractual commitments can be particularly effective to secure an adequate core of solid waste for the facility.

This technique has been successfully utilized in several waste-to-energy facilties in recent years to assure long-term waste supplies. For example, refuse for the Northeast Massachusetts Resource Recovery Project located in North Andover, Massachusetts is delivered to the facility by 22 municipalities whom have signed 20-year "put-or pay" agreements for waste disposal and eight commercial haulers whom have signed private hauler agreements. These "put-or pay" agreements require each community to deliver a guaranteed annual tonnage which can be adjusted yearly within certain limitations. These communities are assessed penalties for shortfalls or excesses below or above their contractual guarantees. Private haulers are also assessed penalties if their deliveries are below or above their contractual guarantee. In this way, the waste-to-energy facility can capture adequate quantities of solid waste to meet its financial commitments to the local communities, the facility operator, and the bondholders.

Economic Incentives for Waste Stream Control

Waste stream control can also be achieved by local government through economic incentives. For example, the operator of a waste-to-energy facility can attract solid waste from both public and private collectors by charging a zero or lower tipping fee than alternate disposal methods such as sanitary landfills. In this case, private haulers would be attracted to the facility since they would have no economic incentive to dispose of their solid waste at less convenient sanitary landfills elsewhere.

In order to accomplish this type of economic control over solid waste for its waste-to-energy facility, the community must be willing to subsidize the loss of project revenues with funds from some other source, such as from the general fund, a user fee, or a tax. For example, a user fee for solid waste disposal can be established for different residential, commerical, and industrial accounts whereby the proceeds from this fee could be used to offset the zero or artifically low tipping fee at the waste-to-energy facility. Some communities have also used revenue from property or other local government taxes to subsidize the tipping fee at their waste-to-energy facilities. Use of property taxes, however, may be viewed by the investment community as a general obligation of the community and could result in a lowering of its bond rating.

REFERENCES

1. William F. Cosulich Associates, The Integration of Energy and Material Recovery in the Essex County Solid Waste Management Program. Belleview, New Jersey: Essex County Division of Solid Waste Management (1983).

2. Felago, Richard T. Waste Stream Assurance: Key to Resource Recovery. Solid Wastes Management July (1982).

3. Franklin, William E., Majorie A. Franklin and Robert Hunt, Waste

Paper: The Future of a Resource 1980-2000. Prairie Village, Kansas: Franklin Associates, Ltd. (1982).

4. Kovacs, William L. Flow Control: An Unnecessary Constitutional Conflict in Managing Solid Waste. Environmental Analyst June: 3-7 (1982).

5. Personal communication from Charles Citrin, Sparber, Shevin, Rosen, Shapo and Heilbronner on September 17, 1982.

6. Snyder, David L., Antitrust Law and Solid Waste Management: The Municipal Perscpective. In: Proceedings of the Governmental Refuse Collection and Disposal Association, August 16, 1982 in Dallas, Texas, Washington: Governmental Refuse Collection and Disposal Association (1982).

7. Spiegel, David R. Local Governments and the Terror of Antitrust. American Bar Association Journal 69: 163-166 (1983).

- 6 -
Selecting the Facility Site

INTRODUCTION

The siting of a major public facility, especially a waste-to-energy project, is not a simple task, particularly when such facilities often are located in highly developed and environmentally conscious communities. Many technical, environmental, and social (institutional) issues must be considered. The site selection process is a complex process, requiring the project developer to not only identify a site that minimizes adverse environmental impacts and can accommodate the operation of a waste-to-energy facility, but also requires the development of specific site evaluation criteria that have a reasonable chance of public acceptance. To achieve this latter objective, the site evaluation criteria so devised must be well documented and carried out in a uniform and consistent manner.

This chapter describes a generic site selection process which can serve as a model for siting efforts of waste-to-energy projects and avoid some of the major siting pitfalls. The method described can allow project developers to identify feasible sites; to eliminate the less suitable ones; and to recommend the best site(s) in a detailed and objective way. Furthermore, these siting methods can help enable communities to win public support for such sites, which is the key to the successful implementation of any waste-to-energy project.

THE SITE SELECTION PROCESS

The site selection process for a waste-to-energy project requires several stages of analysis (Figure 6-1). These steps are in essence screening stages within the selection process by progressively narrowing the criteria for analysis and evaluating more detailed data.

Evaluation Criteria

The first step in the site selection process is to identify and document the criteria. It is necessary that all of the criteria important to the siting of a facility are given balanced consideration. In order to maintain clarity in this effort and to provide a uniform method of reviewing and screening sites, the following three broad categories of site evaluation criteria have been found extremely useful for many projects:

o Technical Considerations;

o Environmental Considerations; and

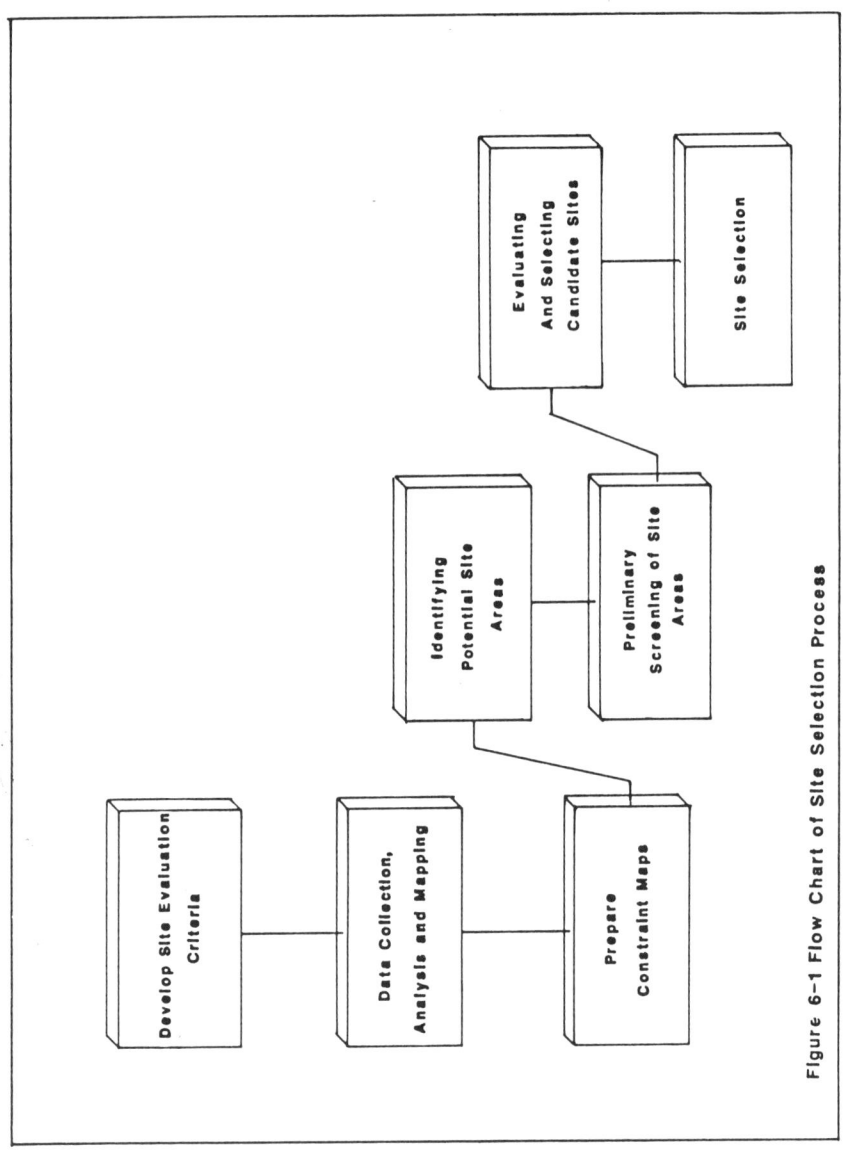

Figure 6-1 Flow Chart of Site Selection Process

o Social (Institutional) Considerations

By structuring the evaluation this way, project developers can analyze proposed facility sites consistently and ensure that critical issues are not overlooked.

Each major division can then be subdivided into smaller evaluation criteria for more specific, detailed appraisals. The criteria under each broad category varies from one area to another based on local situations. Some typical criteria are:

o Technical Considerations

-- Site drainage
-- Foundation suitability
-- Size and shape of site
-- Accessibility (e.g., highway, railroad, barge access, etc.)
-- Location
-- Utilities

o Environmental Considerations

-- Air quality
-- Water quality
-- Biological resources (e.g., fauna and flora)

o Social (Institutional) Considerations

-- Surrounding land uses
-- Permitting considerations
-- Land ownership
-- Cultural resources

By applying these criteria, candidate sites can be identified and a preferred site selected. The specific features of some of the major evaluation criterion are briefly described in the following sections.

Technical Considerations

Site Drainage. Site drainage is an important design consideration for a waste-to-energy facility. While the buildings, roadways, and ancillary facilities can be protected by natural or artificial features to assure protection against flooding, surface runoff must be directed into nearby watercourses to be carried offsite or retained onsite. Small watercourses can often be rerouted and stormwater detention/storage basins can be developed based upon the design requirements of the local government entity. Constructing these improvements usually results in increased development costs for a site. In addition, flood control ordinances in some communities may prohibit construction of facilities in floodplains. Locating facilities in such areas may also increase the cost of insuring the waste-to-energy project, if flood insurance is available.

Foundation Suitability. Waste-to-energy facilities generally require large and complex buildings due to the fact that the equipment such as boilers, generators, and air pollution control devices are unusually heavy. This necessitates that the site for such facilities have stable soils for construction of foundations. While unstable geological conditions can be overcome through the design and construction of more complex foundations, such conditions could preclude the use of an otherwise attractive site because of the additional expenses involved.

Similarly, high groundwater conditions or shallowness of the site to bedrock could also result in more extensive and complex foundation designs. For example, large waste-to-energy facilities are often designed with a storage pit of sufficient capacity to store several days of solid waste fuel. Such storage areas are usually excavated below existing grade to provide the necessary storage volumes. Thus, sites with high groundwater conditions may require that either the entire structure be raised through the use of fill, as an alternative, or a shallow tipping floor be utilized. In either case, such designs may significantly increase the construction and operation of the facility as compared to sites not requiring such innovative foundation designs.

<u>Size and Shape of Site</u>. The size and shape of a site required for a waste-to-energy facility is project specific and could vary from as small as five acres to as large as 50 acres. For example, the size of a site can be dependent on the following key factors: the type of technology used to process the solid waste (e.g., mass burn, RDF, etc.); the quantity of refuse to be processed; and the location and access to the site in relation to neighboring land uses. Consequently, land requirements for waste-to-energy facilities can vary greatly from one community to another. A site surrounded by heavy industrial land uses, for example, may require minimal buffering, thus reducing the acreage needed for the project. However, sites bordering light industrial and residential land uses, or those having flat topography, may require significantly more acreage to provide buffer zones between

the plant and its neighbors. In addition, local traffic and road conditions may also impact the overall size of the site due to the fact that special access road configurations may be required to adequately handle the numbers of vehicles entering and leaving the facility. This could add additional acres to a community's preferred parcel size.

Accessibility. An operating waste-to-energy facility generates significant numbers of vehicles which deliver solid waste, to be processed and hauled away, recovered materials, and residues. Due to the size and numbers of these vehicles, it is preferable that access to a facility site be from major highway and rail systems and not through high-density residential areas to minimize potential traffic congestion and accidents. The ability of the local road network nearest a proposed site to handle the flow of solid waste vehicles safely and efficiently must be examined. Project developers should evaluate such factors as: road widths; structural capabilities of roadways and bridges; weight limits; height restrictions; speed limits; and grades. The purpose is to determine whether existing roadways can safely carry an increased vehicle traffic load. The cost of providing these services can be useful information to assist in the ranking of candidate sites.

Location. The location of a site for a waste-to-energy facility is an important evaluation factor. Such facilities should be located within reasonable distances to the solid waste collection area; the energy sales market; needed utilities such as electricity; cooling and boiler makeup water; and the sanitary landfill which will accept the ash residues or bypass waste.

The ideal location for a waste-to-energy facility is a site located close to the center of the waste generation centroid, adjacent to a sanitary landfill and energy customer, and within the local government's utility service area. It is unlikely that sites meeting all such requirements for waste-to-energy projects can be found. Consequently, project developers often must evaluate the degree to which these requirements are met by the sites under analysis.

Foremost among the requirements for a waste-to-energy facility, is its proximity to a proposed electricity or steam customer. The costs of constructing electrical transmission lines are often prohibitive for most waste-to-energy projects. Thus, the most economical locations for a facility are sites near large electrical transmission lines or existing electric utility substations. In the case of facilities proposing to export steam to an industrial or institutional customer, sites located near an existing steam loop or the customer itself are preferable to minimize the costs of expensive steam distribution lines and their associated equipment.

Another major requirement for a waste-to-energy facility is its location to the existing and future solid waste collection area of the municipality. The closer the facility is to the solid waste generation center, the lower the cost will be to transport the solid waste to the facility. Similarly, sites located in a line between the center of solid waste generation and the sanitary landfill, which will accept the plant's ash residue, have a higher comparative advantage since haul costs of the ash residue can be minimized. However, waste-to-energy processes greatly reduce the volume of the community's solid waste by a factor of nearly 90%, so it is generally more economical to locate the facility near the center of the community's solid waste generation rather than closer to the ash residue landfill. Unfortunately, for many communities, the center of solid waste generation is usually found in their most developed and populated areas making siting in these areas difficult.

Utilities. Waste-to-energy facilities can consume substantial quantities of electric power, water, and natural gas, and require telephone, wastewater and emergency services such as fire and police protection and ambulance services. Sites located where such utility services are already available are preferable to those where these services must be provided at considerable expense to the project. Some utilities such as telephone service are readily available in most communities and can be easily extended, while other utilities such as water and wastewater service may be unavailable in some communities.

Electricity, water, and wastewater service are generally the major utility service needs of waste-to-energy facilities. Such facilities can consume substantial amounts of electric power principally at times of plant startup and outages. Electric service can usually be provided to waste-to-energy plants at reasonable expense in most communities. An electric transmission line or a substation must be located nearby to deliver the facility's energy output.

A sanitary discharge system is a requirement for the proper operation of a waste-to-energy facility. Liquid wastes usually result from many plant operations such as: boiler blow-down; water pretreatment for the boiler; and the normal sanitary requirements of the plant work force. The community's sewer system is the preferable discharge system for most sites. However, there may be instances where such service may be unavailable to handle the flow of the plant because of prior commitments. This may require that project developers would have to consider onsite treatment and discharge, thereby increasing the potential cost of the facility.

A source of water for cooling and process needs is another major utility required for a waste-to-energy plant. Potable water from a public system or an onsite well is usually necessary for normal sanitary needs. Non-potable or recovered water from wastewater treatment plants or nearby rivers can be used for other facility operations such as evaporative cooling, boiler feedwater makeup, and fire protection. The cost of providing this water service to the plant either results from extending existing water lines or drilling new onsite wells. This can impact the desirability of one site versus another requiring minimal expenditures for water service.

Environmental Considerations

Air Quality. The impact of a waste-to-energy plant at a particular site on the local or regional air quality is an important consideration in site selection. Waste-to-energy facilities incorporate some form of combustion process which results in various gaseous and solid emissions to the atmosphere. Good combustion control and the addition of air pollution control equipment, such as electrostatic precipitators, bag houses, and acid gas scrubbers, will help minimize the overall air pollution potential of a proposed facility, although there will still be some quantity of air emissions which could degrade the existing local and regional air quality. Areas designated by regulatory officials as not meeting existing standards for specified air pollutants will generally require more expensive air pollution control equipment than plants located on sites in areas designated as attaining these regulatory standards.

Computer modeling of the area's air quality and meteorological data can go a long way in helping evaluate the potential impact of the facility upon ambient air quality. Such models assist in determining whether the plant can meet minimum regulatory air emission standards with respect to a particular site's specific configuration. In addition, knowledge of an area's atmospheric flow characteristics can also help predict whether normal wind patterns will assist in dissipating the plant's emissions away from sensitive human receptors.

<u>Water Quality</u>. Water quality is also a very important issue in the siting of waste-to-energy facilities. Waste-to-energy facilities utilize significant quantities of water for cooling and process needs. Project developers should consider the impact of the eventual

disposal of liquid wastes from a waste-to-energy facility upon the water quality of nearby bodies of water and groundwater aquifers. Some states have recently considered promulgating stringent regulations restricting the development of certain land areas, located near designated high-quality waters, for construction of certain public works projects such as wastewater treatment plants and solid waste facilities.

Biological Resources. There are a number of unique flora and fauna species that are protected by federal, state, and local regulations. It is important during the initial screening of sites for a waste-to-energy facility to identify the habitats of these threatened or endangered species to ensure that these areas be avoided for development.

Social Considerations

Surrounding Land Uses. The compatibility of a waste-to-energy facility with its surrounding land uses is an important consideration in siting. An operating waste-to-energy facility can exhibit all the indications of an industrial-type of activity. It can generate significant volumes of truck traffic; has a potential to emit noise, odors, and dust; and may because of its building height and configuration suggest the visual appearance of a heavy industry. This is not to suggest that these potential impacts could not be overcome through the use of appropriate mitigative measures. The visual appearance of such facilities, for example, can be made compatible with many land uses through the judicious use of landscaping, buffer zones, and architectural materials such as glass, brick and colored metal panels.

Early determination of land use incompatibility can eliminate significant project delays at later phases of project implementation. For example, areas near airports require special attention since the Federal Aviation Administration (FAA) regulations limit construction (particularly height) in or near airport runway approaches. Distance from airport runways and elevation of ground surface should be documented to determine whether a stack might interfere with aircraft navigation. Building height restrictions are in place in many communities around all the airport facilities, often through strict zoning regulations.

<u>Permitting Considerations</u>. The number of permits and the length of time needed to acquire them for a waste-to-energy facility can be an important factor in the successful implementation of a project. Permitting delays due to the complexity of the regulatory process can impact a project's financial success. Although some permits will generally be required regardless of site location, there are other permits which are applicable based upon specific site conditions. At the outset of a project, it is critical that both the potential number of permits and estimated length of time required to obtain those permits be evaluated for the specific candidate sites. Under this criterion, sites that potentially require the least number of permits would be preferred as compared to those sites requiring a greater number of permits. Consequently, sites that do not contain environmentally sensitive lands which are protected under current

regulations would have greater likelihood of permitting success.

Land Ownership. Land ownership is an important factor in determining the availability and ease of obtaining a site for a waste-to-energy facility. Land parcels under the control of governments are preferable over privately-owned lands because there is less likelihood of acquisition delays. Many governments, however, do not have parcels that can be dedicated for use in waste-to-energy projects. In this case, privately-owned lands, which have only one owner, should be preferred over sites having multiple owners due to the increased ease and speed of land acquisition.

Cultural Resources. Cultural resources include such items as: archaeological areas; historic sites; and scenic landmarks. The construction of a waste-to-energy facility on or near sites having cultural significance can have both direct and indirect effects. Direct impacts can occur as a result of the actual construction and operation of a facility. Cultural resources can also be indirectly impacted if the presence of the waste-to-energy facility affects their use.

SITE SCREENING PROCESS

Once the evaluation criteria have been developed, and qualitative or quantitative ratings assigned to each, project developers can utilize them to screen potential sites. While there are many ways by which this site selection process can be undertaken, ideally the process includes several stages of analysis. That is, there is a progressive narrowing of criteria. The overall site selection process discussed below has worked well in the siting of waste-to-energy facilities and can be adapted to a specific project.

Stage 1: Data Collection and Analysis

In order to apply the siting criteria in the screening process, data need to be collected and compiled to relate these abstract facts to specific places. If for example, flooding is a key concern to a community, then data needs to be collected which can detail the location and nature of flood-prone areas in the area under study. Many federal, state, local and private agencies normally collect such data as part of their normal activities. Meetings should be held with these entities to ensure that the data being supplied is adequate for the community's needs.

Stage 2: Preparation of Constraint Maps

Once the data required for the site screening process are collected, they should be illustrated in a series of maps which can help assist in further data analysis. Many waste-to-energy siting studies have found it useful to portray the data in a series of constraint maps which relate similar criteria together on a single map format. For example, maps containing data on site criteria that may severely restrict development of waste-to-energy projects, can be overlayed to produce a single map which would show those areas with the most restrictive conditions for development.

Some of the criteria so portrayed may have such a significant impact upon development that the location of a waste-to-energy facility would be unsuitable, and these areas would not merit any further consideration. There are other criteria that may adversely impact the development of a site for a waste-to-energy facility unless costly mitigative measures are incorporated in the project design. Some examples of these exclusion site criteria may be the following:

- Air quality non-attainment;
- Wetlands;
- Airport restricted zones;
- Flooding susceptibility;
- Historic sites;
- Environmentally sensitive areas;
- Prime or unique agricultural areas;
- State or local recreational areas;
- Developed lands; and
- Unstable geological conditions (e.g., sinkholes, earthquake zones, etc.)

Similarly, maps showing areas reflecting factors favoring the location of a waste-to-energy plant could also be prepared. Criteria which may favor the location of a waste-to-energy facility in an area could include the following:

- Solid waste generation centroids;
- Location near major highways and interchanges;
- Energy market areas;
- Areas near wastewater treatment plants;
- Water service areas; and

o Areas with industrial, light-industrial or commercial land uses.

Stage 3: Identifying Potential Site Areas

The constraint maps prepared in stage two can be used by study participants to both eliminate areas for further consideration and to suggest general areas within a community which favor the development of a waste-to-energy facility. Potential site areas can then be identified for more detailed onsite analysis.

Stage 4: Preliminary Screening of Site Areas

The next major stage in site screening is to evaluate the potential site areas which have been identified. This usually requires an onsite investigation to determine the actual site conditions at the time of the study. This onsite analysis can often eliminate some sites from further consideration because their actual conditions at the time of study may be different from those indicated on the maps which may have been completed at an earlier date. Onsite analysis is often critical in rapidly-developing areas since sites often shown on maps as vacant land may already have been developed. Such field investigation can also aid in identifying more specific sites for further analysis within the large, undefined areas shown on the maps. It is important that these studies be well-documented through the use of a survey form to assure opposition groups that the study was undertaken in an objective manner.

Stage 5: Evaluating and Selecting Candidate Sites

Based upon the results obtained in stage four, a list of potential sites can be constructed from which candidate sites may be evaluated. At this point in the site selection process, it is useful to obtain additional data on the sites, such as the number of separate parcels of land for each site, their ownership, and land prices. The sites should be revisited to verify the initial field investigation conducted in stage four and to obtain any needed additional site data.

It is at this time that the sites under consideration can be rated using the siting criteria which had been developed for the study. To illustrate this procedure, each site either could be assigned a quantitative or qualitative rating for the technical, environmental, and social siting criteria. As shown in Table 6-1, for example, each site could be assigned a qualitative rating of "good", "average", or "below average", or some similar rating scheme. Thus, assuming that the three criteria categories have equal importance, the overall rating for a particular site would be a composite of its individual ratings for technical, environmental, and social criteria.

In order to more precisely reflect the relative value of each siting criteria a weighting system can be developed. Each individual criterion or criteria category (e.g., technical, environmental, and social) can be assigned a particular weighting to indicate its relative importance. For example, technical considerations could be assigned a weighting of three; environmental

TABLE 6-1
EXAMPLE OF WEIGHTED RATINGS FOR WASTE-TO-ENERGY SITES

Sites	Ratings			
	Technical	Environmental	Social	Composite
A	A	G	G	G-A
B	G	A	A	G-A
C	G	G	G	G
D	G	A	A	G-A
E	G	G	G	G
F	G	A	A	G-A
G	G	A	G	G
H	A	G	A	A
I	BA	BA	BA	BA
J	A	G	G	G-A
K	G	BA	A	A
L	BA	G	A	A
M	A	A	G	A

Sources: References 1 and 4

[a] G = Good; A = Average; and BA = Below Average.
[b] Technical considerations given a weighting of three; environmental considerations a weighting of two; and social considerations a weighting of one.

considerations a weighting of two; and social considerations a weighting of one. Such weightings can be developed through discussions with the community's decision-makers, civic and environmental interest groups, and the general public. By utilizing this weighting procedure, the siting team can narrow its search to relatively few candidate sites.

The first task of the team in evaluating such sites is a more detailed review of the data collected earlier, and the collection and analysis of additional data. The detailed evaluation of these remaining sites could include the following steps:

- o Meetings and correspondence with property owners and other landowners adjacent to the candidate sites;

- o Meetings and correspondence with a number of federal, state, local and private agencies to gather and evaluate data;

- o Detailed site visits, including discussions with nearby residential civic associations, and officials of public institutions such as schools, hospitals, and emergency services that might be required by the waste-to-energy facility;

- o Completing a preliminary air quality impact analysis at each site, including modeling of emissions and detailed permitting requirements;

o Undertaking a preliminary traffic impact assessment at each site, including modeling potential vehicle flows at different roadway segments at each site;

o Analysis of noise impacts, including measurements of existing ambient noise levels and assessments of predicted noise impacts at each site;

o Assessment of required mitigative measures to address any potential aesthetic and visual impacts at each site;

o A geological investigation at each site, including several subsurface borings; and

o Comparative cost analysis of each site, including relative differences in transporting, constructing and facility operating costs.

Stage 6: Site Selection

The site evaluation criteria from stages one through five form the basis for selection of the candidate sites. In many cases, there are few sharp differences in the overall ratings among the remaining sites themselves. Consequently, in order to make more detailed distinctions among these sites, many studies have found it necessary to regroup their evaluation categories to address key project issues such as: compatability with adjacent land uses; environmental impact; and comparative project costs; among others. These factors can then be assigned numerical values for each site. As illustrated in Table 6-2, a "one" could signify, for example, the most valuable ranking, and a "four" could signify the least valuable ranking. These factors could then be added together and the site with the overall lowest numerical value could then be recommended for acquisition.

Land Use Compatibility. Evaluation of land use compatibility can play an important role in the final selection of a project site. This evaluation issue could include those siting criteria which may have impacts upon a proposed waste-to-energy facility with its surrounding land uses. A technical criterion, such as accessibility, could be used to determine the impact of the plant's traffic flows upon neighboring land uses. Social criteria, such as surrounding land uses, land ownership, and cultural resources, could be utilized to assess the effect of the plant on existing or proposed land uses. Permitting considerations are another criterion that would reflect the need for zoning and comprehensive plan modifications to assure the site's compatibility with adjacent or surrounding land uses. An environmental criterion, such as air quality, could be used to reflect concerns with possible facility impacts upon surrounding areas.

Environmental Impact. Major environmental constraints, which would prohibit the development of a waste-to-energy facility, would have been

TABLE 6-2
EXAMPLE OF THE COMPARATIVE RANKING OF FOUR CANDIDATE
WASTE-TO-ENERGY SITES

Selection Parameters	Site F	Site H	Site B	Site P
Compatibility With Adjacent Land Uses	2	3	1	4
Environmental Impact	2	3	4	1
Comparative Costs	1	3	4	2

Sources: References 1 and 4

considered in earlier stages of the siting analysis. The environmental impact review of the candidate sites in this stage, however, can focus on their relative differences. For example, the following environmental categories can be examined in detail: air quality; water quality; site drainage; and biological resources. The use of such categories can allow the evaluation to focus on the most critical environmental issues associated with a waste-to-energy facility, and on environmental features that may differ from site to site. For example, the results of the air quality impact analysis can provide additional input into the analysis by helping predict the need for additional air pollution control devices or an increase in stack height due to potential downdrafts of air pollutants at a particular site. Technical criteria, such as site drainage and accessibility, could be utilized to reflect potential impacts of flooding and traffic flow, respectively, upon the local environment.

Comparative Costs. All candidate sites can be compared for costs. Thus, differences between the sites can be considered, assuming that identical facilities could be constructed at each site. This comparison could include technical criteria such as: site drainage; geological conditions; utilities; and accessibility. These would reflect relative construction costs of the plant at each site. The relative differences in the costs of hauling solid waste to each site from the center of the solid waste collection area could be assessed using the criterion of location. Environmental criteria, such as air quality and water quality, could also be considered.

On the basis of air quality modeling, the relative cost of installing the necessary air pollution control equipment required for a facility could be calculated for each site. Permitting considerations can also be applied under this category, both in terms of the estimated costs of obtaining permits, and also in delays resulting from the acquisition of the necessary permits.

REFERENCES

1. Camp, Dresser and McKee, Inc., Evaluation and Selection of a Resource Recovery Site For Hillsborough County, Florida, Tampa: Hillsborough County, Florida (1982).

2. Hauser, Robert, Marc J. Rogoff, and Paul J. Stoller, Site Selection of a Resource Recovery Facility in Hillsborough County, Florida. In: Proceedings of the Governmental Refuse Collection and Disposal Association 23rd Annual Interational Seminar, August 23-30. Silver Spring, MD: Governmental Refuse Collection and Disposal Association (1985).

3. Rogers, John W., Resource Recovery Facility Siting. In: Proceedings of Fourth Annual Resource Recovery Conference, March 28-29 in Washington, D.C., Washington: National Resource Recovery Association (1985).

4. Rogoff, Marc J., Paul J. Stoller and Robert Hauser, How to Site a Resource Recovery Facility. World Wastes, July (1984).

5. Rogoff, Marc J., Warren N. Smith and Patricia Berry, Siting of a Resource Recovery Facility: Hillsborough County, Florida. In: <u>Proceedings of the Fourth Annual Resource Recovery Conference</u>, March 28-29 in Washington, D.C., Washington: National Resource Recovery Association (1985).

6. Wade, David, Siting and the Thermal Energy Market. In: <u>Proceedings of Third Annual Resource Recovery Conference</u>, March 28-29 in Washington, D.C., Washington: National Resource Recovery Association (1984).

- 7 -
Energy and Materials Markets

INTRODUCTION

The primary objectives of any proposed waste-to-energy project are to dispose of solid waste in an economically efficient and environmentally acceptable manner. Although the derived energy from the combustion of municipal solid waste in a waste-to-energy project is a secondary benefit, the revenues gained through the sale of electricity, steam, or recovered materials help offset the cost of solid waste disposal. By turning solid waste into useful energy and materials, waste-to-energy projects are able to compete against other waste disposal options which are much less capital intensive. The financial viability of any waste-to-energy facility, therefore, is in large measure dependent upon its ability to sell energy to a long-term customer. This chapter will explore some of the issues project developers should consider in investigating the energy and materials markets for any waste-to-energy project.

ENERGY MARKETS

Steam

Prior to the Arab Oil Embargo in 1973, most incinerators in the United States recovered little or no thermal energy, except for internal plant needs. The worldwide increase in energy costs after 1973 made energy recovery from solid waste a viable option. Many American cities, which once had established steam distribution systems for their downtown areas, but had abandoned them in the era of cheap energy, began to seriously consider the redevelopment of these systems. The steam produced by incinerating solid waste was considered as an energy source for steam distribution networks of single or multiple customers.

Typically, steam from waste-to-energy facilities in the United States is transported to industries at temperatures ranging from 250 degrees F to 750 degrees F and pressures ranging from 150 to 650 pounds per square inch. Pipelines transporting steam are short, rarely longer than two to three miles in order to minimize energy losses and capital investment. When steam is transported it experiences a drop in pressure and temperature. Steam losses from such systems are estimated to be about 30 to 40 pounds per square inch per mile of transmission line. Consequently, the location of the energy customer is an important factor in the siting of a waste-to-energy facility.

Steam is the most desirable form of energy from solid waste, if an appropriate long-term steam customer can be located. The revenues, which the developer of a waste-to-energy facility will receive from any customer, however, is dependent on the price of customer's existing energy service; the customer's steam needs, on a continuous basis; and the guaranteed reliability of the steam produced by the waste-to-energy facility. The highest prices for steam are received by those facilities which can continuously produce steam on a schedule that matches the customer. Most facilities are unable to produce a guaranteed steam supply, and thus produce a limited steam supply requiring the customer to have a backup steam source. Consequently, the value of this steam to the customer is lower.

In summary, a waste-to-energy developer should consider the following critical factors in the marketing of steam:

- o Siting: The waste-to-energy facility should be close in proximity to the steam customer(s);

- o Price: The steam produced must be delivered at a price competitive with the customer's existing primary source;

- o Operating Schedule: Steam must be produced at a level matching the needs of the customer;

- o Availability of Fuel: Enough waste must be available to produce the contracted quantity of steam;

o Cost of Connecting Facilities: Who should pay the cost of installing steam lines, acquiring right-of-ways, and installing condensate return lines between the facility and the steam customer(s);

o Steam Quality: Who is responsibile to keep contaminants out of the steam lines and meet the customer(s) steam quality; and

o Contracts: Can the steam customer sign a long-term put-or-pay contract; what will happen if he defaults.

Electricity

A long-term steam customer for a waste-to-energy facility may be unavailable in many areas of the United States. In those instances, the conversion of solid waste into electricity may be a feasible alternative. The electricity generated by the facility can be utilized for internal needs of the project with surplus power being available for sale to utilities; transferred ("wheeled") over the transmission of one utility to another; or made available for use by other governmental operations.

Prior to the enactment of the Public Utilities Regulatory Policies Act of 1978, commonly known as PURPA, waste-to-energy facilities seeking to sell their surplus power to public or investor-owned utilities faced significant obstacles. For example, many utilities refused to purchase electricity from such producers or offered to purchase power at low rates. In addition, some utilities charged discriminatorily high rates for back-up electric service.

In order to conserve oil and natural gas, the U.S. Congress enacted PURPA with the objective of removing obstacles for facilities which generated electricity from alternative fuels such as solid waste, wood waste and biomass. Under Section 201 of PURPA, electric utilities are required to provide electric service to "qualifying facilities" at rates which are "just and reasonable, in the public interest, and which do not discriminate against cogenerators and small power producers." Qualifying facilities or QFs are defined under the Act under two categories--small power production facilities and cogeneration facilities. A small power production facility is one which produces less than 80 megawatts of electricity from biomass, waste, or renewable resource fuel constituting more than 50 percent of the total energy source. Non-renewable fuel sources, such as oil, natural gas, or coal, cannot constitute more than 25 percent of the total energy input of the QF during any year. In addition, an electric utility or its subsidiary may not hold more than 50 percent of the equity interest of a QF.

Cogeneration facilities are somewhat different from small power producers in that they can produce and sell electricity and other forms of energy such as hot water or steam. Topping-cycle facilities are those which produce electric power first and then thermal energy. Bottoming-cycle facilities are those which produce thermal energy first and then electric energy. For such facilities to be deemed qualifying facilities, PURPA requires that they meet certain specific operating and efficiency standards and the same ownership criteria as small power producers.

Certification as a qualifying facility may take two forms under PURPA. The cogenerator or small power producer can merely "notify" the Federal Energy Regulatory Commission (FERC), which administers PURPA, that their facility is a QF. This option make sense for those facilities whose characteristics do not vary dramatically from those prescribed under PURPA. The other option available for those projects is to submit an application to FERC requesting the agency to certify the qualifying status of the facility. In December, 1985, FERC instituted a filing fee of $7,800 for processing such certifications.

Under administrative rules promulgated by FERC to implement PURPA, utilities must purchase electric energy and generating capacity made available by QFs. The electric power purchased must be at rates reflecting the cost that the purchasing utility can avoid to produce or purchase from other utilities as a result of obtaining energy and capacity from the QF. Hence, the terms "avoided energy" and "avoided capacity". The FERC rules require that utilities must furnish data on avoided energy and capacity so that QFs are able to estimate such revenues.

While the FERC rules set the general policy requiring utilities to purchase electricity from QFs, the state public utility/service commissions are charged with enforcing this right. As one might imagine, each state commission has implemented the FERC rules quite differently in light of conditions existing in its own state. Consequently, the electric revenues and the type of permitted contractual relationship between utilities and a QF varies from state to state. For example, some states like New York have established a statutory six cent per kilowatt hour (Kwh) floor payment by electric utilities to cogenerators and small power producers, while other states like California have established complex formulas for computing avoided cost buyback rates. Several states have even allowed QFs to receive higher avoided capacity payments than utilities would incur in building new generating capacity because the power generated by cogenerators and small power producers has been deemed more reliable than central generating stations.

Negotiating power sales contracts has been both a complex and expensive process for developers of waste-to-energy projects. Such contracts often form an integral part of the financing of these projects since electric revenues constitute a major portion of their long-term revenue bases. In response to the need to streamline the negotiating process between utilities and QFs, the increasing trend for state public service commissions has been produce "standard contract offers" establishing a tariff which all electric utilities in their state must offer to virtually any qualifying facility. These standard contracts include provisions on subjects such as: regulatory risk; security requirements for early payments by utilities; production reliability; and interconnection requirements. For example, qualifying facilities in Florida, which execute a

"Standard Offer Contract" can receive avoided energy and early capacity payments equal to those costs incurred by a hypothetical "Statewide Avoided Unit". QFs are required by the Florida Public Service Commission, however, to secure these early capacity payments through a letter of credit, escrow account, or similar cash equivalent security, or in the case of a governmental project, by its full-faith and credit or other similar security.

MATERIALS MARKETS

The recovery of reusable materials from the solid waste stream has had a long history in the United States. These materials have traditionally been recycled from the waste stream, particularly paper and metals, when market conditions are favorable. Unfortunately, however, at times when markets for recovered materials are poor, due to fluctuations in the business cycle, these materials are unrecovered and must be disposed of by local government in sanitary landfills and waste-to-energy facilities. What is normally not recognized by waste-to-energy project developers is that materials recovery programs are compatible with the implementation of waste-to-energy projects and compliment each other. Given the abrasive nature of glass, for example, its removal would be a welcome relief for processing plant maintenance crews. Some of the advantages that a viable materials recovery program can offer include:

o Recovering glass and metals
  from the waste stream will
  provide a higher quality

fuel for a waste-to-energy facility in most cases;

o Recycling removes materials which might be damaging to a waste-to-energy system;

o Materials recovery reduces the quantity of materials which would ordinarily be landfilled (ash residue, and non-processible wastes) in a waste-to-energy facility;

o A viable materials recovery system can help reduce the overall size and costs of a waste-to-energy facility;

o Conserves essential natural resources, including energy; and

o Helps extend the life of landfills.

The following subsections will discuss the types of materials which can be recovered and recycled from municipal solid waste, the technologies available for recovering these materials, and the types of markets typically available for such materials.

Materials in the municipal solid waste stream generally targeted for recycling include the following:

o Ferrous metals;
o Non-ferrous metals;
o Glass;
o Paper products; and
o Film plastics

Ferrous Metals

Ferrous metals, which are available in the solid waste streams of most communities, are primarily in the form of tin-coated steel cans and white goods. These metallic materials represent about five to ten % by weight of the municipal waste stream in many U.S. communities. They can be sorted out manually in source separation programs, or by magnets through front-end separation processes in RDF systems which can typically recover 80 to 85 percent of ferrous metals in the incoming waste. Mass-burn systems can also employ permanent or electromagnetic separators and trommels to remove ferrous metals from the ash residue, but the incineration process produces a slag containing contaminants. Thus, these ferrous materials have significantly less value to scrap dealers, and the price quoted is substantially lower than if such materials were recovered before incineration.

The markets and price received for ferrous metals vary with the existing industry demand, the quality of the material recovered,.and the cost of transporting these materials to the market. Many waste-to-energy projects recovering ferrous metals from the waste stream have had difficulty negotiating long-term contracts due to the cyclical nature of the ferrous market.

Non-Ferrous Metals

Aluminum is the only non-ferrous metal that can be economically recovered from municipal solid waste. It usually represents less than one percent by weight of the typical municipal solid waste stream. Other non-ferrous metals, such as lead and copper, are usually recycled by industries producing these materials.

The aluminum industry has purchased aluminum from municipal source separation programs in the form of discarded beverage containers which are reasonably clean and dry and free of glass, steel, paper and other contaminants. This market continues to be strong due to the high cost of producing aluminum cans. The demand has increased from about 26,500 tons in 1972 to over 500,000 tons in 1984. High-tech reverse vending machines placed outdoors in supermarket parking lots or lobbies are a recent trend.

Aluminum can also be recovered from the solid waste stream either through manual or mechanical means. Manual separation is highly labor intensive, but is said to result in the recovery of about 45 percent of the aluminum in the waste stream. There are also a number of mechanical and media separation techniques that have been developed using trommels, disc screens, and other devices making use of the differences in the specific gravity of aluminum. Experience to date with the later equipment has not been entirely successful.

Glass

Municipal solid waste contains an average of four to eight % glass by weight. Glass can be recovered from the municipal solid waste stream, either as broken or non-broken glass, and also sorted into different colors of glass such as flint (clear), amber, or green. Color separation enhances the market value of the glass material. Mixed glass material, termed "cullet", has had poor marketability due to the glass industry's need for glass materials that are clean and free from refractory particles for the manufacture of clear bottles.

In recent years, the glass industry has increased its interest in recovered glass materials. Many manufacturers are recycling their cullet which they generate internally. The use of such waste glass in their production processes allows them to reduce their fuel consumption since cullet is easier to melt than virgin raw materials. In order to maintain a desired minimum cullet percentage in their bottling facilities, they are increasingly becoming buyers of cullet in the open market.

Paper Products

Waste paper products, such as newsprint, corrugated paper (cardboard), and high grade office paper (computer print-out paper, tab cards, and ledger paper), typically represent about 30 to 40 percent by weight of the municipal solid waste stream. These materials must be processed through source separation programs since mixing them in the solid waste stream renders them unacceptable for recycling. Separation of newsprint, corrugated paper or boxboard, and high-grade office paper has proven to be economically feasible in many areas, although the prices for these materials have fluctuated dramatically depending on market conditions.

Film Plastics

The average municipal solid waste stream contains about one to four percent plastic materials. During the last decade, the percentage of plastic materials in the waste stream has increased dramatically because of the surge in its use for a variety of household products, particularly in the field of plastic film packaging. Unfortunately, recycling of these products is currently nonexistent. Some facilities in Europe are utilizing mechanical processes to recover the film plastics from the waste stream and manufacture plastic products such as trash bags and pipes.

REFERENCES

1. Scaramelli, Alfred B., Energy Market: Key to Project Planning. <u>Solid Wastes Management</u> April: (1982).

2. Davitian, Harry, Power Sales Contracts for Resource Recovery Projects. In: <u>Proceedings of the Fifth Annual Resource Recovery Conference</u> in Washington, D.C. on March 28-30, 1986. Washington: National Resource Recovery Association (1986).

3. U.S. Environmental Protection Agency, <u>Guides For Municipal Officials: Markets</u>, Washington: U.S. Environmental Protection Agency (1976). SW- 157.3.

- 8 -
Permitting of Waste-to-Energy Facilities

INTRODUCTION

Long before construction activities for a waste-to-energy facility can begin, project participants must have received all the necessary federal, state and local regulatory approvals for construction and operation. The environmental permitting process is potentially the most time-consuming and controversial step on the road to project implementation. This is due in part to the extensive data needed on such projects which often must be submitted to governmental agencies in the form of detailed permit applications and environmental impact statements. In addition, the permit approval process for waste-to-energy projects in most jurisdictions across the United States is unfortunately fragmented among multiple federal, state, and local agencies. This provides opposition groups several forums to attack and delay the approval process of such projects both in administrative hearings and through judicial review. Consequently, environmental permitting can become a tortuous and frustrating experience for many communities if they have not developed a strategy to expedite the permitting process.

This chapter will provide a general overview of the types of permits which may be required for a waste-to-energy facility. The reader should be forewarned that the number and specific permits required depends on project specific and site specific circumstances.

FEDERAL PERMITS

Under existing environmental laws and regulations, there are many federal agencies which may review the potential impacts of a waste-to-energy facility upon the environment (Table 8-1). These agencies may also require that the community submit a specific permit application to a federal agency for their review and approval before construction and/or operation of its waste-to-energy facility can commence. Discussed in the paragraphs below is a description of these major federal permits.

Air Quality

Air quality is one of the most important aspects of the permitting of a waste-to-energy facility. Such facilities produce a wide range of compounds, including the traditionally regulated criteria pollutants (e.g., sulfur dioxide, particulates, carbon monoxide, nitrogen dioxide, lead, and ozone), and non-criteria pollutants for which standards have not been set by environmental agencies. A great deal of the public outcry against waste-to-energy facilities has been focused on the air quality issue.

TABLE 8-1
POTENTIAL MAJOR FEDERAL PERMITS FOR WASTE-TO-ENERGY PROJECTS

Act or Regulation	Regulators	Actions Reviewed	Specific Permits
Clean Air Act	EPA State[a]	Air emissions of regulated pollutants	Federal PSD Permit
Clean Water Act	EPA State[a]	Discharge of water pollutants to ground or surface waters	NPDES
Rivers and Harbors Act Clean Water Act Marine Protection Research and Sanctuaries Act	COE	Dredging and filling in navigable waters and wetlands of U.S. and disposal of spoil material	Dredge and Fill Permits
Federal Aviation Act	FAA	Construction which may impact air navigation	Notice of Proposed Construction
Endangered Species Act	DOI	Potential impact on habitat of rare, protected, or endangered species	None
Public Utility Regulatory Policies Act	FERC	Qualification as QF or small power producer	QF Status
National Environmental Policy Act	EPA	Federal funding or major Federal action	EIS

[a] Permitting authority has been delegated by EPA to some states.

The Federal Clean Air Act (42 U.S.C. 7401 et seq., as amended), and the regulations promulgated by the U.S. Environmental Protection Agency (EPA), form the legal basis for permitting air emissions from waste-to-energy facilities. EPA has delegated this permitting authority to the environmental regulatory agencies of several states.

Currently, the U.S. Congress, through enactment of the Clean Air Act, has required that the EPA ensure that facilities generating air emissions comply with the following standards:

- o National Ambient Air Quality Standards (NAAQS);

- o National Emission Standards for Hazardous Air Pollutants (NESHAP);

- o New Source Performance Standards (NSPS);

- o Prevention of Significant Deterioration of Air Quality (PSD);

- o Non-Attainment Area New Source Review Requirements

National Ambient Air Quality Standards (NAAQS). Pursuant to authority under the Clean Air Act, the EPA has established National Ambient Air Quality Standards (NAAQS) for certain regulated air pollutants known as "criteria pollutants" (Table 8-2). Primary NAAQS and secondary NAAQS have been promulgated for these criteria pollutants. Primary NAAQS are those standards deemed necessary by EPA to protect the public health.

TABLE 8-2
NATIONAL AMBIENT AIR QUALITY STANDARDS

Pollutant	Interval	Standard Levels ($\mu g/m^3$) Primary	Standard Levels ($\mu g/m^3$) Secondary
Sulfur Oxides	Annual[a]	80	--
	24-hour	365	--
	3-hour	--	1300
Particulates	Annual[b]	75	60
	24-hour	260	150
Carbon Monoxide	8-hour	10	10
	1-hour	40	40
Nitrogen Dioxide	Annual[a]	100	100
	1-hour	240	240
Lead ($\mu g/m^3$)	3-month	1.5	--
Ozone	3-hour	160	160

[a] Arithmetic mean.
[b] Geometric mean.

Secondary NAAQS are those standards for criteria pollutants which are established to protect the public welfare. States are required by the Clean Air Act to ensure that primary and secondary NAAQS are not exceeded through the development of State Implementation Plans (SIPs) which specifically determine how each state will achieve the primary and secondary NAAQS. These plans define those geographic areas of each state which attain ("attainment areas") or do not attain the NAAQS ("non-attainment").

National Emissions Standards for Hazardous Air Pollutants (NESHAP). The Clean Air Act also empowers the EPA to regulate the emissions of hazardous or toxic air pollutants from waste-to-energy facilities. Pursuant to this authority, EPA has published a list of hazardous air pollutants and established emission standards for each pollutant. Pollutants currently listed by EPA are asbestos, beryllium, mercury and vinyl chloride.

Prevention of Significant Deterioration of Air Quality (PSD). The NAAQS for the criteria air pollutants serve as a baseline for air quality in the United States. Areas which are in compliance with these standards are classified as being "attainment" air quality regions. Areas which are in violation of these standards are classified as being "non-attainment" air quality regions. Consequently, the siting of a waste-to-energy facility must take into account whether the region is either attainment or non-attainment for these standards since this existing air quality condition may impact the types of control technology necessary for the facility; whether air pollution offsets from adjacent plants may be required; or whether a permit may be issued at all.

Since 1977, EPA has promulgated air regulations to prevent the significant deterioration of air quality in regions where NAAQS is in attainment. The intent of such regulations is to prevent the degradation of the nation's existing air quality by the operation of new air pollution sources.

The "Prevention of Significant Deterioration of Air Quality (PSD)" regulations impose strict requirements on the amount of additional air emissions above ambient concentrations, called "PSD increments", which may be introduced in any area in the United States. Once these increments are consumed by industry for criteria pollutant air emissions, further emissions of these pollutants are not allowed by new facilities without a corresponding reduction of emissions from existing facilities in the region. As Table 8-3 indicates, air quality increments for SO_2 and TSP which the EPA has established for the three different air quality areas in the nation. Class I areas have the smallest increments and include national parks, wilderness areas, and other areas where air quality deterioration has been deemed undesirable as a matter of public policy. Class II areas have substantially more allowable increments and allow for moderate, controlled growth. Class III areas allow additional air emissions pollution up to the secondary NAAQS.

A waste-to-energy facility desiring to locate in an air quality attainment area, and which has the potential to emit more than 100 tons per year of any

TABLE 8-3
PSD AIR QUALITY INCREMENTS BY AIR QUALITY AREAS

Pollutant	Maximum Allowable Increase (micrograms per cubic meter)		
	Class I	Class II	Class III
Particulates:			
Annual geometric mean	5	19	37
24-hour maximum	10	37	75
Sulfur Dioxide:			
Annual arithmetic mean	2	20	40
24-hour maximum	5	91	182
3-hour maximum	25	512	700

pollutant, is considered by EPA to be a major stationary source, and therefore subject to a PSD review. This requires that the permit applicant must present data to EPA or the state agency based on the expected emissions of the proposed facility. If any pollutant is emitted at a rate greater than a significant emission rate (Table 8-4), it is subject to PSD review. This may require that the following information be submitted with the permit application: a Best Available Control Technology (BACT) analysis; an air quality impact analysis; and possibly an ambient air quality monitoring program.

BACT analysis is done on a case-by-case basis, although EPA disseminates information on control technology for different types of stationary sources. The aim of an BACT analysis is to determine the best available air pollution control or state-of-the-art technology to reduce the emissions of a specific pollutant, taking into account energy, environmental, and economic impacts, and other costs. EPA disseminates BACT determinations on a national basis through the BACT Clearinghouse.

Once the BACT analysis is completed, the next step in the PSD review is for the applicant to complete an analysis of ambient air quality in the vicinity of the proposed facility. In some cases, an applicant will be required to submit monitoring data which had been gathered continuously over a 365-day period, although shorter periods are common. Applicants may be exempt from these monitoring requirements if current representative data are available or ambient impacts can be shown to be below

TABLE 8-4

AIR POLLUTANTS REQUIRING PSD REVIEW

Pollutant	Threshold Emmission Rate (tons/year)
Carbon Monoxide	100
Nitrogen Oxides	40
Sulfur Dioxide	40
Hydrocarbons	40
Particulates	25
Hydrogen Sulfide	10
Total Reduced Sulfur	10
Sulfuric Acid Mist	7
Fluorides	3
Vinyl Chloride	1
Lead	0.6
Mercury	0.1
Asbestos	0.007
Beryllium	0.0004

SOURCE: 40 CFR 52.21 (1984)

de minimus, or insignificant, monitoring concentrations as listed in Table 8-4.

The air quality impact analysis prepared by the applicant is done through air quality simulation modeling which takes into account ambient air quality, climate, meteorology, terrain, and other factors near the proposed site. The intent of such modeling is to predict the ground level impacts on the sources on the PSD increments. In addition to this required modeling, applicants may be required to do additional impact estimates such as the impact on a Class I area, or the impact on neighboring attainment and non-attainment areas.

New Source Review Standards Analysis. In addition to a PSD review, a waste-to-energy facility locating in an non-attainment area is subject to a New Source Review Standards Analysis (NSRSA). These regulations require that a proposed facility, if it emits 100 tons per year or more of any pollutant for which the NAAQS is already exceeded in the air quality region, must meet standards more stringent than the BACT determination. This requirement, called the "Lowest Achieveable Emission Rate" (LAER), is defined as the most severe emission limitation which is contained in the implementation plan of any state for a waste-to-energy facility, unless the applicant can demonstrate that such limitations are not technically achievable regardless of cost.

Water Quality

Federal agencies have a key permitting role in the area of water quality. They have been delegated by

Congress regulatory authority over the discharge of pollutants into the "waters of the United States" and dredging and filling activities. Each of these potential permit requirements for waste-to-energy facilities are discussed below.

National Pollution Discharge Elimination System Permits. The Clean Water Act (P.L. 92-217 and P.L. 95-576) was enacted by Congress to "restore and maintain the chemical, physical and biological integrity of the nation's waters." In furtherance of this objective, Congress authorized EPA to develop a nationwide permit program (the National Pollutant Discharge Elimination System-NPDES) which imposes limitations on pollutant discharges in the waters of the United States. The term "waters of the United States" is broadly defined to include: navigable waters; tributaries of navigable waters; interstate waters; intrastate lakes, rivers and streams used by interstate travelers for recreation; industries involved in interstate commerce; and wetlands. Under the Clean Water Act, every public or private waste-to-energy facility that discharges waste waters, including process and washdown water, must have a NPDES permit.

EPA is the issuing agency for all NPDES permits. Some states, however, have been authorized by the EPA to administer the NPDES program. This permit generally sets forth discharge limitations for certain pollutants, requires monitoring and periodic reports to the EPA or state agency on compliance, and schedules of compliance, and so forth. These determinations all organized into a draft permit which is advertised for public comment. After a 30 day comment period, the EPA or issuing agency holds a public hearing if there is a significant degree of public interest.

Dredge and Fill Permits. Under federal law, the U.S. Army Corps of Engineers has been granted regulatory authority in the areas of navigable water and wetlands. Pursuant to the Rivers and Harbors Act of 1899, the Corps of Engineers is responsible for permitting all dredge and fill activities, dock construction, and pier and bulkhead improvements in navigable waters. Consequently, if construction of a waste-to-energy facility required such shoreline improvements, then the Corps of Engineers would have to be involved in the permitting of these modifications.

The Corps of Engineers is also responsible for the regulation of dredge and fill activities in wetlands under Section 404 of the Clean Water Act Amendments of 1977. Thus, should construction of a waste-to-energy facility require the dredging and/or filling of a wetland, then a Corps of Engineers permit will also be required.

In some states, joint permit application forms and public notices have been developed in order to expedite the permit approval process. This application is also reviewed by other federal agencies such as the Fish and Wildlife Service and National Marine Fisheries Service.

Federal Aviation Administration Permit

Under federal law, the Federal Aviation Administration (FAA) has responsibility to assess the impact of structures which may be a hazard to air traffic. With respect to waste-to-energy facilities, the FAA must approve the construction of the stack of each facility, taking into account its height, elevation, and location to nearby airports. Applicants must file a "Notice of Proposed Construction or Alteration" notifying the FAA at least 30 days prior to the start of the initial construction. In addition, each sponsor who undertakes this construction activity must notify the agency at least 48 hours prior to the start of construction, and also within five days after the construction (stack erection) reaches its greatest height.

The FAA has the authority, after review of the applicant's notice, to issue either a "no hazard" or "hazard" ruling. If a no hazard ruling is made, construction can begin. Typically, the agency mandates the requirements for marking and lighting of the stack structure. If a hazard ruling is made, the FAA has no authority to prevent construction but the project may become uninsurable.

STATE AND LOCAL PERMITTING REQUIREMENTS

In addition to the possible federal environmental permits required for a waste-to-energy facility, permits to construct and operate such a facility must be obtained from state and local regulatory authorities. These requirements can include: approvals from these government entities for construction and operation of a new source of air emissions; construction and operation of a waste-to-energy facility; construction of a stormwater discharge facility; dredging and filling of the proposed site location in waters of the state; or permit approvals for the construction and operation of an ash residue landfill in conjunction with the waste-to-energy facility. Additional permits may be required from regional authorities with jurisdiction, for example, over groundwater and surface water withdrawals needs for operation of the facility.

For example, in the State of Florida, the Department of Environmental Regulation (DER) is lead agency responsible for the review and approval of waste-to-energy projects. Pursuant to state statutes, DER has the authority to regulate air emissions, discharges of stormwater, disposal of solid waste, dredging and filling of state waters, and wastewater discharges. These state regulatory requirements are similar to those found in most other states.

Permits for the construction and operation of a waste-to-energy facility may also be required from local regulatory authorities such as: planning commissions; environmental protection boards or commissions; and airport authorities. These permit applications may include: land use and zoning variances; building permits; boiler inspection reviews; and similar approvals. Opposition for waste-to-energy projects may surface during these public hearings required for such permits. Project sponsors need to anticipate that public opposition may occur. This requires that a complete record be kept of such public proceedings should a legal challenge be brought against the project by opponents.

KEYS FOR A SUCCESSFUL PERMITTING STRATEGY

The permitting of a waste-to-energy facility can become a tedious, costly, and time-consuming element in the implementation of a project. Attention to the potential environmental regulatory requirements at the early stages of project development can help avoid these possible pitfalls. This requires that project sponsors develop a comprehensive permitting strategy. Some successful projects have found that the following elements are necessary:

- Assemble an experienced permitting team of consultants as early as possible to include engineering, legal and financial talent with in-depth working knowledge of the regulations;

- This team should meet with all regulatory agencies at the earliest possible time to conduct scoping meetings to help eliminate unnecessary permit requirements, multiple submittals and hearings;

- All submittals to agencies should be carefully reviewed by the permitting team;

- Understandings between the project sponsor and the permitting agencies should be confirmed in writing to avoid misunderstandings at later points in the permitting process;

o Work closely with the agencies to develop their trust and assistance; and

o Establish a strong political and public information program to effectively respond to public concerns about the project.

THE DIOXIN ISSUE

No current discussion regarding the environmental permitting of waste-to-energy facilities would be complete without a brief overview of the dioxin issue--currently one of the most controversial issues in the waste-to-energy industry. "Dioxin" is a generic term for a family of 75 polychlorinated organic compounds known as benzo-p-dioxins (PCDDs) which are thought to be generated in waste-to-energy facilities through the combustion of municipal solid waste. Closely associated with dioxins are "furans" which are a family of 135 similar polychlorinated compounds, dibenzofurans (PCDFs), which differ chemically from dioxin in that its two benzene rings are linked by one, rather than two oxygen atoms. The organic compounds within each family differ from each other by the number of attached chlorine atoms and their location on the benzene rings.

The public concerns over the emissions of these man-made compounds from waste-to-energy facilities is centered on their potential impacts on human health. Dioxins and furans are some of the most toxic substances known to man, 500 times more potent than strychnine and 10,000 times more potent that cyanide as demonstrated by laboratory tests on small animals. Research on the human health impacts of inhaling large doses of dioxins and furans is currently in very early stages, since the only information available is from accidental releases of these compounds from plants involved in the manufacture of agricultural and industrial chemicals. A rigorous examination of these industrial accidents has been difficult because the prior medical histories of those exposed to these compounds is often unavailable. In any case, it appears that the symptom most frequently associated with exposure to dioxin is chloracne, a skin condition.

Much of what is known about the health impact effects of exposures to dioxins and furans has been learned through animal experiments using 2,3,7,8 TCDD. This type of dioxin compound is perhaps the most toxic. These animal studies have conclusively shown that this substance is extremely toxic to some animals, such as the guinea pig and rat, even in small amounts. On the other hand, hamsters with the same body weight as the guinea pigs and rats are 5,000 more resistant to TCDD. Existing data also show similar inconsistencies with respect to laboratory dogs and monkeys. Consequently, it is extremely difficult to extrapolate from these laboratory results in order to make predictions or risk assessments on human health. Human studies cannot be undertaken because of the high level of risk involved.

There has been a great deal of controversy among scientists on how and where dioxins are generated and/or destroyed in the process of incineration. There currently is a lack of scientific consensus on a wide range of issues concerning dioxin genesis, control, and health risk. The formation of TCDDs and PCDDs in the combustion of municipal solid waste are not yet thoroughly understood, although various theories have been proposed:

- o TCDDs and PCDDs are trace components of municipal solid waste are thus are not formed during the combustion process;

- o TCDDs and PCDDs are formed from precursors such as PCBs, chloro-phenols, and other similar organic materials; and

- o PCDDs are formed <u>de novo</u> from unrelated materials, such as plastics, petroleum products, and hydocarbons.

Research studies are continuing as of this date under the sponsorship of federal and state regulatory agencies, as well as private professional organizations. Results from these studies should make it possible for regulatory agencies to make better informed decisions on the possible health and environmental effect of dioxin from waste-to-energy facilities. In the interim, some state regulatory authorities can be expected to set some upper limits for dioxin emissions from these facilities.

REFERENCES

1. California Air Resources Board, <u>Air Pollution Control at Resource Recovery Facilities</u>, Sacramento:

California Air Resources Board (1984).

2. Commoner, Barry, Thomas Webster and Karen Shapiro, Environmental Levels and Health Effects of PCDDs and PCDFs. Paper presented at the 50th International Symposium on Chlorinated Dioxins and Related Compounds, September 16-19, 1985 in Bayreuth, West Germany.

3. Dee, David S., Permitting For the Hillsborough County Solid Waste Energy Recovery Facility. In: <u>Proceedings of the First Annual International Resource Recovery Symposium</u>, held in Tampa, Florida on February 18-21, 1986, Silver Spring, MD: Governmental Refuse Collection and Disposal Association (1986).

4. Fred C. Hart Associates, Inc., <u>Assessment of Potential Public Health Impacts Associated With Predicted Emissions of Polychlorinated Dibenzo-Dioxins and Polychlorinated Dibenzo-Furans From the Brooklyn Navy Yard Resource Recovery Facility</u>, New York: New York City Department of Sanitation (1984).

5. Henningson, Durham and Richardson, Inc., <u>Risk Assessment For Trace Element and Organic Emissions</u>, Santa Barbara: Henningson, Durham and Richardson, Inc. (1984).

6. Hichey, Maurice D., <u>Resource Recovery and Solid Waste Management in Norway, Sweden, Denmark, and Germany: Lessons for New York</u>,

Albany: New York State Legislative Commission on Solid Waste Management (1985).

7. Kemp, Clinton C. <u>Notes on Polychloro Dibenzo Dioxins and Polychloro Dibenzo Furans in Connection With Waste-to-Energy Plants</u>, Houston: Browning-Ferris Industries (1983).

8. Niessen, Walter R., Production of Polychlorinated Dibenzo-p-Dioxins (PCDD) and Dibenzofurans (PCDF) From Resource Recovery Facilities Part II. In: <u>Proceedings of the Fourth International Offshore Mechanics and Arctic Engineering Symposium - Volume I</u>, New York: American Society of Mechanical Engineers (1984), pp 358-376.

9. Shaub, Walter M., <u>Technical Issues Concerned With PCDD and PCDF Formation and Destruction in MSW Fired Incinerators</u>, Gaithersburg: U.S. Department of Commerce (1984). NBSIR 84-2975.

10. State of Florida, <u>An Overview of Dioxin in Florida</u>, Tallahassee: Florida Department of Environmental Regulation (1986).

11. Snyder, David L., Permits Strategy for Resource Recovery Projects. In: <u>Proceedings of the Third Annual Resource Recovery Conference</u>, March 29-30 in Washington, D.C., Washington: National Resource Recovery Association (1984).

12. Steisel, Norman, Environmental Problem-Solving and Environmental

Law. Speech presented to the New York State Bar Association Section on Environmental Law at Pace University, White Plains, New York on December 13, 1985.

- 9 -
Procurement of Waste-to-Energy Systems

INTRODUCTION

The procurement of a waste-to-energy system by a community is one of the final steps on the long road of project implementation. Prior to embarking on this final path of system procurement, each community would have already addressed the difficult decisions, as discussed in previous chapters, which are necessary to implement the project. The recent history of waste-to-energy project implementation in the United States has shown that many projects have suffered serious setbacks, if not abandonment, because their organizers had not addressed these decisions. Oftentimes, these decisions were not made even after the procurement process had already begun.

Once these critical project decisions are made, the procurement process for a waste-to-energy facility can begin. Communities have wide latitude in the approaches and procedures they can utilize in procuring waste-to-energy systems, although many state statutes and local charters and ordinances may limit the choices available to some communities. Since a waste-to-energy facility is usually the most capital intensive and complex public works project attempted by most communities, great care must be taken at the outset by the community's legal advisors to develop a procurement process which both meets the community's needs, and also follows state and local statutes.

This chapter will discuss some of the general approaches and procedures, which may be used by local government, to procure waste-to-energy systems. The procurement approach determines the manner by which engineering, design, construction, start-up, and operations services are acquired. The procurement procedure also dictates the method by which such services can be acquired under the legal mandates of the government. A case study of the procurement process used in Hillsborough County, Florida will serve to illustrate how a community might develop a Request-for-Proposal, evaluate bidder responses, and conduct final contract negotiations.

PROCUREMENT APPROACHES

The process selected by a community to procure a waste-to-energy system can be approached in three different ways:

o Architect/Engineer (A/E) Approach;

o Turnkey Approach; and

o Full-Service Approach

Table 9-1 identifies some of the project

TABLE 9-1
TYPICAL WASTE-TO-ENERGY PROJECT RESPONSIBILITIES BY
PROCUREMENT APPROACH

Project Responsibilities	Procurement Approaches		
	A/E	Turnkey	Full Service
Planning	G or E	G or E	G or E
Preparation of Plant Specifications	E	C	C
Plant Design	E	C	C
Construction Supervision	E	C	C
Construction and Equipment Installation	C	C	C
Startup	G or C	G or C	C
Operation	G or C	G or C	C
Ownership	G	G	G or C

Source: Reference 6

[a] G = Government; E = Engineer; and C = Contractor.

responsibilities which are typically undertaken by government, contractors, and the architect/engineer in waste-to-energy projects.

Architect/Engineer (A/E) Approach

The architect/engineer approach is the one most generally used for large public benefit projects in the United States. For a typical project, the government agency first procures the services of an architectural or engineering firm which is then responsible for preparing the plans or system specifications and certain design elements of the particular project. These materials are then distributed by the community as part of an advertised competitive-bid process.

Once the winning bidder is selected, the architectural or engineering firm, which prepared the project specifications and plans, or another similar type of firm, is oftentimes retained by the community to monitor the construction of the project, prepare operating manuals, and assist in the start-up and acceptance testing of the project. Once the project is accepted by the community as meeting contractual obligations, the community is then responsible for either operating the project itself or contracting out this responsibility to a private firm. This approach requires multiple contracts between the local government and its architect/engineer, general construction contractor, and equipment suppliers.

The conventional A/E approach has been modified slightly by some local governments in their procurement of waste-to-energy projects. Unlike the traditional A/E approach of bidding each individual piece of equipment for the project, the entire process line and turbine generator equipment, commonly referred to as the "chute-to-stack" in mass burn facilities, is bid as a single package. The A/E firm is still responsible for designing the ancillary facilities. This approach has the advantage of minimizing the number of potential vendors that local government must deal with, while providing a mechanism for government for sharing the risk of project performance with a private entity. In recent years, three waste-to-energy facilities in the United States have been procurred using this approach: Olmsted County, Minnesota, City of Commerce, California, and Norfolk, Virginia.

Turnkey Approach

The turnkey approach differs from the A/E approach in that a single private firm is responsible for the design, construction, and startup of the project. Under this approach, the turnkey contractor is responsible for acquiring the necessary equipment and supplies for the project as well as ensuring that the architectural or engineering design work is prepared. Once startup and acceptance testing is completed, however, the turnkey contractor turns over the responsibility for operating the project to the community. This approach may make perfect sense for those communities which have the technical capabilities to operate waste-to-energy facilities. Other communities desiring to retain the option of public operation may modify this approach by negotiating a short-term initial operating agreement which would require the contractor to solve design and operational problems after acceptance testing before turning over full operation to a municipal workforce.

Full-Service Approach

A modification of the turnkey approach is for a community to assign total responsibility to a private firm over the conduct of the project, including the design, construction, startup, testing, operation, and possibly, the ownership and project financing responsibilities. The full-service approach can enable a community to acquire the services of a waste-to-energy facility without making the community responsible for its long-term, day-to-day operation and maintenance.

PROCEDURES FOR CONDUCTING THE PROCUREMENT PROCESS

A basic tenet of procurement policy in the United States is that fair and open competition minimizes the costs of goods and services to a community. Competition is believed by many to improve the quality of the goods and services purchased, increase the choices available, and foster innovation among vendors. In addition to these goals, there is a general public belief that competition reduces favoritism and inspires greater confidence in the governmental procurement system.

Waste-to-energy systems are highly complex and sophisticated public benefit projects. Procurement procedures that may be appropriate for obtaining public goods and services at the lowest possible cost for a traditional public benefit project may not be desirable in the case of a waste-to-energy facility. Discussed in the paragraphs below are four general procurement methods which have been recommended by the American Bar Association. Not all these procedures are applicable for use with the procurement approaches available for waste-to-energy facilities, as previously discussed. For example, the competitive sealed bidding procedure is not appropriate in turnkey or full-service procurement processes because factors like vendor experience, technology, guarantees, and financial capability, are considered along with price to select the winning contractor. Table 9-2 provides a summary of the relationship between procurement approaches and procedures.

Competitive Sealed Bidding

Competitive sealed bidding or formal advertising is the standard procurement contracting method used by most communities for acquiring equipment and services for public benefit projects. Typically, the community or its consultant prepares the Invitation-for-Bid (IFB) which is used to solicit bids for the specific project. Contractors prepare their bids based upon the specifications described in this document with no discussions with the purchasing officials of the community. These sealed bids are usually opened in public at a time and place stated in the IFB. Generally, the bid is promptly awarded to the bidder submitting the lowest price, provided this bidder can satisfy the contractual standards of the community and is responsive with all the terms of the IFB.

Competitive bidding can be used in the procurement of a waste-to-energy

TABLE 9-2
APPLICABILITY OF PROCUREMENT APPROACHES AND PROCEDURES
FOR WASTE-TO-ENERGY PROJECTS

Procurement Procedures	Procurement Approaches		
	A/E	Turnkey	Full Service
Competitive Sealed Bidding	Appropriate	Not Appropriate	Not Appropriate
Competitive Negotiation	Possible	Appropriate	Appropriate
Multiple Step Process	Not Appropriate	Appropriate	Appropriate
Sole Source	Not Appropriate	Appropriate	Appropriate

Source: Reference 6

facility, provided that the following project conditions are present:

- o Definitive specifications have already been prepared for the waste-to-energy facility;

- o Price or some other factor is the only criterion of choice; and

- o More than one response to the IFB can be expected.

Multiple-Step or Simultaneous Negotiations Method

This multiple-step procedure, sometimes called "simultaneous negotiations", has been utilized by the federal government for the procurement of advanced engineering and operational systems for which detailed specifications had not yet been prepared by the purchasing agency. These multiple-step procedures incorporate components of competitive bidding and competitive negotiation. At the outset of the procurement, an RFP is advertised requesting the submittal of an unpriced technical proposal which is evaluated by a selection committee upon receipt. Some governments have used an RFQ as a method to limit the number of proposers who receive the RFP. At this stage, discussions are held with proposers to determine the type of materials and services offered, and to allow proposers to make changes in their proposals to enable them to be responsive to the solicitation document. These discussions are concluded with the offering agency requesting a "best and final" offer on a common cutoff date by vendors. These final offers are then reviewed and evaluated. Thereafter, the process is similar to the competitive bidding process in that the proposer whose bid is most responsible and complete is awarded the overall contract.

In summary, the competitive bidding process is most useful when specifications are complete for a waste-to-energy project and price alone is the only factor for choosing the best overall vendor.

Competitive Negotiation

Competitive negotiation is quite different than the competitive-sealed bidding process used for conventional public works projects. In many communities, this process has been reserved for the procurement of specialized services, such as architectural, planning and engineering services, where discussions between the offering agency and vendor can help clarify the reasonableness of proposals. Consequently, competitive negotiation differs from the competitive sealed bidding process in two basic respects:

- o Price is not the only criterion of choice. The use of other factors helps the offering agency to determine which proposal is most advantageous to the offering agency; and

- o Discussions are held with vendors after the submittal of proposals to determine important technical, financial and management interrelationships of each proposal.

Competitive negotiation has been that most common procedure utilized by communities to procure complex public works projects, such as waste-to-energy systems. Under this procedure, the community solicits proposals, not bids, with documents termed Request-for-Qualifications (RFQ) and Request-for-Proposals (RFP). In some cases, communities release a single procurement package, which contains both the RFQ and RFP, to speed up the procurement process. Typically, this procurement process can begin with the offering agency advertising the release of a RFQ to system vendors. This document contains some comments from the community on system performance, procurement schedule, desired technology, and financing requirements. In most cases the RFQ requires vendors, who desire to prepare proposals on the project, to submit very specific technical, managerial and financial qualifications regarding their undertaking of the project. For example, respondents to the RFQ may be asked to provide technical data on their performance capabilities, such as operational data from commercial-sized reference plants utilizing their technology.

Once these submittals are analyzed by the community, vendors deemed qualified to undertake the project receive a copy of the RFP package which contains detailed performance specifications desired by the community for the waste-to-energy project. This document specifically describes criteria, which will be used by the community, to evaluate the proposals. Shortly after the RFP is released, many communities schedule a pre-proposal conference to help brief the potential proposers on information contained within the RFP; provide clarifications to proposer's questions; and review the procurement schedule. If as a result of this conference, and subsequent written questions submitted by proposers, changes in the RFP are determined to be advisable by the offering agency, amendments to the RFP can be released. Since the complexity of these changes may affect the responsiveness of all proposals, many communities have changed their deadline for proposal submission.

Following receipt of the proposals, and their detailed technical evaluation, the community may begin discussions with either the top-ranked proposer or enter into simultaneous negotiations with two or more of the top-ranked proposers. This option of simultaneous negotiation must be specifically described by the community in the RFP document. These discussions may cover any items contained within the proposals, as long as the information derived by one proposal is not disclosed to another competing proposer. Following these discussions, each proposer is offered the opportunity to revise its proposal and submit a "best and final offer" to the community. After evaluation of these final offers, an award of the procurement is announced by the community.

Sole-Source Negotiation

A sole-source, negotiated procurement involves no competition, and is usually restricted to cases where the time factor for procurement is extremely critical or where there is only one source of supply for the desired waste-to-energy technology. Under a sole source procurement, a vendor either submits an unsolicited proposal or one in response to an RFP. Upon receipt of this proposal the community evaluates the proposal and negotiates the terms and conditions of the award of the project. Sole-source procurement of waste-to-energy systems is often open to criticism on the basis of favoritism and high cost due to the absence of market competition.

PREPARING THE REQUEST FOR PROPOSAL

The Request-for-Proposal (RFP) document is critical to the success of a community's procurement process for a waste-to-energy facility. This document is designed to acquaint potential proposers with the specific technical, financial, institutional, and contractual aspects of the project. Clearly, it must be able to communicate the proposal requirements of the issuing agency to potential proposers. As such, this document must be carefully drafted by the community to eliminate ambiguities which might result in some proposers not responding to the RFP. The RFP, therefore, is the community's statement of its goals for the project to the waste-to-energy industry. Each RFP must be tailored specifically to the needs of the particular community. Consequently, a "standard RFP" does not exist.

This section describes some of the information that an issuing agency should include in a RFP at a minimum for a waste-to-energy facility. An RFP prepared for Hillsborough County, Florida is used as an illustrative example.

Format of an RFP

The RFP used by Hillsborough County, Florida had the following seven major divisions:

- Introductory Materials: This section provided a listing of project team members and included the table of contents;

- General Information to Proposers: This section provided proposers with an overview of the County and the project;

- Instructions for Proposal Preparation and Submission: this section provided instructions for the preparation of proposals and how and when they should be submitted;

- Technical Requirements: The purpose of this section was to inform proposers of the requirements, constraints, and technical conditions for the design, construction, and operation of the project;

- Proposal Forms: This section provided proposal forms to be completed by each proposer;

- Draft Agreements: This section provided draft design and construction and operations and management agreements which required comment by each proposer; and o Evaluation, Selection and Negotiation Process: This section described the process used by the County to evaluate each proposal and how the negotiations process would be undertaken. This section provided a listing of consultants and staff, such as engineers, investment bankers, and special counsels, who had responsibility in the format and preparation of the community's RFP. Such information is critical since the proposer's decision to respond to an RFP can be affected by the quality of the community's consulting team. Following this listing, a complete table of contents was provided for the reader's easy access to the document.

General Information to Proposers

This section was designed to provide proposers with an overview of the Hillsborough County project. Introductory remarks were made indicating that the county was requesting a full-service vendor to design, construct, startup, test, operate and maintain the waste-to-energy facility for 20 years, but that the County would reserve ownership rights. Furthermore, this introductory subsection indicated the location of the facility site, its permitting status, and the system size (1,200 tons per day) boiler (three) and generator (one) configuration.

This introductory subsection was followed by other subsections which provided background information on

Hillsborough County such as:

- Location, size and population;
- Governmental structure;
- Solid waste management system;
- Solid waste quantities;
- Energy market;
- Waste flow control;
- Permitting requirements and responsibilities;
- Site location and description;
- Subsurface conditions;
- Preliminary site plan;
- Utilities; and
- The financing plan including a description of the draft bond ordinance, tax issues, and bond validation.

Instructions for Proposal Preparation and Submission

This section was designed to serve as a guide for preparing and submitting a proposal in response to the RFP. The explanation of the competitive-sealed proposal method of procurement under Hillsborough County purchasing procedures was provided. These introductory remarks were followed by subsections providing general information on such topics as:

- Proposal submission location;
- Proposal submission deadline;
- Addendum and interpretations;
- Proposal preparation expense;
- Signature and authority requirements;
- Openness of procurement process;
- Security bond requirements;
- Errors and omissions by proposers;

o Retention and disposal of proposals;
o Schedule of project events subsequent to release of RFP;
o Pre-proposal conference; and
o Organization of proposal responses.

Hillsborough County required that each copy of each proposal be submitted in the following four physically separate and detachable volumes for evaluation purposes:

o Volume I: Executive Summary
o Volume II: Technical Description
o Volume III: Price Proposal
o Volume IV: Qualifications

The remaining subsections provided detailed instructions for the proposers for preparation of each of these respective volumes. For example, the executive summary was limited to 30 pages, including tables, figures, and illustrations such as the artist's rendition of the Facility.

The technical proposal was required to describe the major design and construction details of the project with specific instructions provided in the RFP on the following required components: a project site plan; process flow description and diagram; project arrangement drawings; major equipment description and specifications; materials of construction; process mass balance diagrams; process energy balance diagrams; process control diagram; project availability analysis; an artist's rendition and landscaping plan; performance guarantees; electrical interconnection plan; overall construction management plan project schedule; project milestone schedule; project startup procedures; project operating plan; personnel requirements; utility utilization; operational performance of project; and environmental compliance.

The price proposal section for each proposal was designed to aid each contractor in summarizing prices and to assist the county in evaluating proposals. This subsection generally described the proposal forms that each respondent was required to complete, as well as the instructions the proposer should follow in commenting on the draft design and construction, and operations and management agreements included with the RFP. As part of the price proposal, the proposers were instructed to state all the conditions and terms in these draft agreements to which exception was taken, using legislative drafting format.

This instructional section concluded with a discussion of the required format of the qualifications proposal. Each proposer was instructed to provide specific details regarding their technical experience or its licensor's (if any) overall experience on similar projects, including actual records for such plants; detailed information on its full-time employees directly involved with the project; and detailed information on major subcontractors. Each proposer was also instructed to provide the following financial data: recent 10K filings with the U.S. Securities and Exchange Commission or audited financial statements for the three previous years; the last quarterly financial reports for the prior two years; a full and complete description of its legal and financial guarantees; a description of its license agreement, if any, in terms of service and guarantees; a copy of any financial document offering any of the firm's financial offerings; credit ratings; and a copy of each firms latest annual report, if any.

Technical Requirements

The purpose of this section was to inform the proposers of the requirements, constraints, and conditions for the design, construction, and operation of the project. To facilitate evaluation of all technical proposals, the RFP instructed the proposers to the utilize design data and information presented with the RFP in the following areas:

- o Project capacity: A minimum of 372,000 tons per year of reference waste with a continuous design rating of 1,200 tons per day; and

- o Reference waste composition, characteristics and ultimate analysis.

The purpose of this section was to specify the minimum design requirements for the project in the following areas of the project:

- o Scale and weigh system;
- o Tipping area;
- o Solid waste storage area;
- o Fire protection system;
- o Fragmentizer;
- o Overhead cranes;
- o Combustion/steam generation units;

- Boiler feedwater and treatment system;
- Residue system;
- Air pollution control systems;
- Flues and stack;
- Power generation system;
- Electrical switchgear;
- Sitework, building and structures including architectural and landscape treatment; and
- Process control and monitoring system.

Proposal Forms

This section included various forms which the proposer was instructed to complete according to the instructions in the RFP and attach to the respective proposal volumes.

Draft Agreements

This section included the draft Design and Construction agreement and Operations and Management Agreement which the proposer was instructed to comment on using the legislative drafting format.

Evaluation, Selection and Negotiation Process

This last section of the RFP described the procedure the County would utilize to review and evaluate each proposal to first determine its completeness and, then, to undertake the evaluation of each of the four volumes of each proposal. This section specifically described the ranking and weighting process to be utilized according to scores in the following areas:

- o Technical evaluation: 20%;
- o Aesthetic and architectural appearance: 10%;
- o Qualification evaluation: 20%;
- o Economic evaluation: 30%; and
- o Agreements response evaluation: 20%.

The section also described the negotiations process to be utilized by the County once proposals were evaluated.

PROPOSAL EVALUATION

The process utilized by a community to evaluate proposals submitted in response to its RFP must be carefully developed to ensure that all proposals are reviewed and evaluated fairly. This evaluation process need not be time-consuming if proper pre-planning is undertaken to: select and train staff who will evaluate such proposals; to develop the necessary comprehensive forms to help evaluators summarize data from each proposal; and to develop and test computer models to analyze economic and financial data derived from the proposals. The following section describes a methodology used by Hillsborough County, Florida to evaluate responses to its RFP for a waste-to-energy system.

Log-In Procedure and Proposal Handling

In accordance with instructions in the RFP, all sealed proposals were to be received by the Hillsborough County's Purchasing Department prior to the opening time prescribed in the RFP and addenda subsequently released. At the time of proposal opening, all proposals received by the county were unsealed, and the following general information was recorded for each: design and construction price, base operating fee per ton, excess operating fee per ton, and whether a bid and performance bond were included with each proposal, as required by the RFP.

Since the RFP required that ten copies of all volumes of the entire proposal, one set was resealed and deposited in the County Clerk's Office for safekeeping. This was done in case questions came up at a later date regarding the accuracy of the remaining proposal sets. The other nine sets of proposals were then transmitted to the County's Department of Solid Waste which was responsible for coordinating the proposal evaluation effort.

Evaluation Committee

A proposal evaluation team for a waste-to-energy project should be composed of team members who are experienced in the areas of engineering, construction management, economics, finance, and law. Prior to the release of the Hillsborough County RFP, an evaluation team was developed which was composed of County representatives from the following key departments: Office of the County Administrator, Department of Solid Waste, Division of Public Works, Division of Fiscal Services, Clerk's Office, and the Office of the County Attorney. This team was assisted by the following members of the County's consulting team: engineering consultant; financial advisor; insurance advisor; investment bankers; and legal advisors.

The following subcommittees were established which were charged with the responsibility of reviewing and evaluating specific components of each proposal for ranking purposes:

- o Technical Evaluation;
- o Aesthetic and Architectural Appearance;
- o Qualifications Evaluation;
- o Economic Evaluation; and
- o Agreements Response Evaluation.

Review of Completeness and Conformance with RFP

Following receipt of the proposals, they were reviewed by the subcommittees with respect to completeness and conformance with the instructions and requirements specifically indicated in the RFP. These requirements were mandated by official actions of the County Commission. As a result of this review, one proposal was deemed incomplete and non-conforming, and another was deemed non-conforming, with respect to the requirements and instructions of the RFP, and neither was evaluated further by the County's Evaluation Committee.

Detailed Evaluation

Each of the remaining four proposals was evaluated using criteria listed in the RFP for evaluation of the technical design, aesthetic and architectural appearance, qualifications, price, and agreements response (Table 9-3). For each of the above categories, a score from one to ten points was assigned to each proposal. The best proposal in each category was assigned the top score of ten points for that category. Each of the other three proposals were awarded a score

TABLE 9-3
CRITERIA USED TO EVALUATE PROPOSALS
HILLSBOROUGH COUNTY, FLORIDA

Technical Evaluation

- Feasibility and operational reliability of equipment and unit processes
- Soundness of plan for project integration of processes and equipment
- Contingency capabilities of proposed system
- Demonstration of ability to comply and maintain compliance with all environmental regulations and permit conditions
- Project expansion capability and ease of expansion
- Safety design features and plan
- Quality of residue produced
- Adaptability of system to technological and regulatory changes
- Management plan and construction schedule
- Operating and maintenance plan

Qualifications Evaluation

- Construction experience and applicability of experience cited
- Operation management experience
- Financial strength
- Credit reports

Aesthetic and Architectural Evaluation

- Completeness of information
- Architectural requirements in RFP
- Overall visual appearance
- Landscaping and site plan

Economic Evaluation

- Net present value of disposal costs

Agreement Response Evaluation

- Level of acceptability of construction and operations agreements
- Level of risk assumed by proposer
- Time frame likely to be required to negotiate
- Proposer's compliance with agreement conditions required to meet financing plan

Source: Reference 4

based upon its comparison to the best proposal within each category.

Following the assignment of points in each of the above categories for each proposal, the score of each of the categories was multiplied by a weighting factor that reflected the County Commission's decision of relative importance of the categories. A weighting factor of one was applied to the aesthetic and architectural appearance category. A weighting factor of two was applied to the technical, qualification, and agreements response evaluation categories. A weighting factor of three was applied to the economic evaluation category, reflecting its perceived importance over the other categories. The total score was determined for each of the four proposals by adding together the total points times the weighting factor in each of the five categories. The maximum potential score was 100 points.

NEGOTIATIONS PROCESS

Based upon the evaluation of the proposals and the ranking process prescribed in the RFP, Hillsborough County staff sought authorization from the County Commission to negotiate final proposed Design and Construction and Operations and Management Agreements with the top-ranked proposer. The County did not exercise its option, available in the RFP, of simultaneous negotiations at the outset with two or more vendors. As an alternative, Hillsborough County staff was authorized to begin negotiations with the first ranked vendor. A time limit of 40 working days was established as the maximum allowable negotiation period. If the negotiations were not completed within this period (which could be terminated any time), the County Commission had the option of instructing staff to extend the negotiation period or enter into negotiations with the next ranked proposer.

Competitive negotiations between a community and a proposer for a waste-to-energy project requires considerable pre-planning on the part of both parties. Each side needs to establish its negotiation objectives, its tactics, and counter proposals to the critical issues. Unfortunately, there is no formal procedure for undertaking negotiations for waste-to-energy projects. This does not mean that negotiations need be time-consuming. With proper pre-planning, the parties can strive to determine the critical areas of disagreement and their relative importance, and develop bargaining strategies for narrowing these differences.

There are three possible outcomes of any negotiation:

- o Win/Lose - One side is elated and the other side is a loser determined to get even;

- o Lose/Lose - Both sides are worse off than before and leave the table with distrust, frustration and hostility; and

- o Win/Win - Both sides are satisfied and get something.

The "win/win" outcome is the most preferable in negotiations for the procurement of a waste-to-energy facility.

Both the community and the proposer leave the negotiating table believing that they were successful in achieving their goals and objectives set at the outset. Agreements entered into under such conditions are much more likely to be respected by either party than under the "win/lose" or "lose/lose" outcomes.

REFERENCES

1. Anonymous, Requests For Proposals in State Government Procurement University of Pennsylvania Law Review 130: 179-215 (1981).

2. Feldstein, Sylvan, Resource Recovery Revenue Bonds: An Analyst's Primer, New York: Merrill Lynch, Pierce, Fenner and Smith (1983).

3. Hayden, John A., The Full-Service Approach to Resource Recovery. Public Works: 68-70 (1983).

4. Hillsborough County, Florida, Request-for-Proposals for a Solid Waste Energy Recovery Facility, Tampa: Hillsborough County, Florida (1984), pp. 6-1 - 6-5.

5. Schoenhofer, Robert F., Michael A. Gagliardo and Harvey W. Gershman, Fast Track Implementation of the Southwest Resource Recovery Facility. In: Proceedings of 1982 National Waste Processing Conference, New York: American Society of Mechanical Engineers (1982), pp 339-350.

6. State of New York, Resource Recovery Procurement: A Guidebook for Community Action, Albany:

New York State Department of Environmental Regulation (1980).

7. Yaffe, Harold J. and Jonathan Wooten, The Development and Financing of the Northeast Massachusetts (NESWC) Resource Recovery Project: A Tale of Twenty-Two Cities and Towns. In: <u>Proceedings of the 1984 National Waste Processing Conference</u>, New York: American Society of Mechanical Engineers (1984), pp. 102-110.

- 10 -
Ownership and Financing of a Waste-to-Energy Facility

INTRODUCTION

A discussion of resource recovery financing must first include a review of project ownership options. In general, the selection of the source of capital to fund the project and the decision as to whether a government entity or private sector party owns the waste-to-energy facility is usually decided together. This chapter will discuss the key factors involved in the selection of an ownership and financing plan for waste-to-energy projects. The ownership of a waste-to-energy project is one of the most important policy decisions which a community must make. Table 10-1 lists some of the major factors which must be considered the choice of public or private ownership.

TABLE 10-1
MAJOR DECISION FACTORS AFFECTING FACILITY OWNERSHIP

Decision Factors	Impacts Under Different Ownership Options	
	Public	Private
Control Over Project	Yes	No
Operating Risks	Low	Lower
Energy and Materials Revenues	Public Owner	Private Owner
Cost to System Ratepayers:		
Initial	Higher than Private Option	Lower than Public Option
Final	Lower than Private Option	Higher than Public Option
Residual Value Entitlement	Yes	No
Property Tax Payments	No	Yes

Sources: References 1 and 3

OWNERSHIP ALTERNATIVES

Public Ownership

The traditional government-operated, government-owned facility for providing public services has been used in most municipalities or counties for schools, roads, water and wastewater plants, and similar facilities. Since this is the traditional approach to financing public works projects, all the legal and institutional mechanisms are usually in place to acquire a waste-to-energy facility in this manner.

Under this ownership alternative, the government entity bears all the operational risks, and expects, in return, to provide a waste disposal service at the lowest operating cost to its ratepayers. However, since local government is not a federal taxpayer, it is unable to realize the available tax savings from such normal business deductions as depreciation or investment tax credits if allowed under the federal tax code. The inability to use these tax benefits commonly results in a higher capital financing cost for publicly-owned, waste-to-energy projects. On the other hand, government would acquire a balance sheet asset which it would retain after the repayment of borrowed capital.

Private Ownership

Private industry can play a major role in the development of waste-to-energy projects. The reason local government may consider private ownership is that the infusion of equity capital will reduce the project capital needs either through the financial impact of the initial private equity contribution, and thus, smaller bond size, or through annual equity contributions during the early years of the project. This is accomplished through project financing combining tax-exempt bonds, and possibly taxable bonds, and private equity, where private owners or third-party investors will obtain federal tax benefits associated with ownership of an industrial facility in addition to the attractive financial return. Equity capital contributions, however, must be structured in order for the private party to be recognized as the facility owners, and thus be eligible for any tax benefits. Local governments, faced with debt and budget limits and unwilling to assume technological risks, have increasingly transferred responsibility to the private sector.

Many waste-to-energy projects implemented in recent years are privately-owned, although the availability of private equity capital in future years will be influenced by changes in federal tax and leasing laws under the recently enacted, Tax Reform Act of 1986 (the "Act"). This federal income tax legislation will have significant impact on the tax benefits once available to non-governmental owners of waste-to-energy facilities. For example, the Act puts into place a somewhat less favorable accelerated cost recovery system for solid waste projects, as well as repealing the use of investment tax credits. Under other provisions of the Act, individuals and closely-held corporations will be unable to deduct their losses from "passive activities", such as limited partnerships, where they do not actively participate on a day-to-day basis. Consequently, private equity capital for waste-to-energy projects from such sources may be somewhat limited in future years.

Perhaps one of the most limiting factors, however, in the development of privately-owned, waste-to-energy projects may be the availability of tax-exempt, industrial development bonds to finance the construction of such facilities. While tax-exempt, industrial revenue bonds would remain available for waste-to-energy facilities, the Act places a statewide cap on the issuance of tax-exempt industrial development bonds for all so-called "private activity bonds" to the greater of $75 per resident or $250 million until December 31, 1987, after which the cap is reduced to $50 per capita or $150 million for each state. Since privately-owned, waste-to-energy facilities would have to compete with many other private uses of such bonds, there is a distinct possibility that private waste-to-energy project developers in some states may be unable to secure a bond allocation for their projects. On the other hand, governmentally-owned facilities would be exempt from this cap.

Special transition rules enacted by Congress for waste-to-energy facilities may allow certain projects to utilize the previously available more favorable tax benefits for privately-owned facilities. Project developers should contact their tax and financial advisors to determine if these transition rules apply to their project.

The major disadvantage of a privately-owned facility is usually perceived to be in the fact that the local government will never own the waste-to-energy facility even though its taxpayers,

through payment of the disposal fees, have helped retire most of the bonds that paid for the construction of the facility. Furthermore, the municipality will have limited control over the facility, except to require periodic tests to demonstrate plant performance guarantees.
Consequently, a privately-owned, privately-operated facility must be analyzed as a long-term provision of a service, not as a purchase of the project.

FINANCING OPTIONS

Waste-to-energy facilities are capital intensive projects. Expenditures for such projects represent a significant, and complex funding problem for any community, requiring a thorough evaluation of several alternative methods of financing. Most communities do not have the available capital out of their general revenue fund to "pay-as-you-go" during construction of the facility. As discussed in the paragraphs below, the capital required to build a waste-to-energy facility may be raised from public sources, private sources, or combinations of public and private sources. Borrowing from such sources results in financing expenses related to the market interest rate on the bonds or rate of return expected by private equity investors.

General Obligation (G.O.) Bonds

General obligation (G.O.) bonds are secured by the government's pledge of its full faith, credit, and taxing power. Although such bonds are secured by a pledge of _ad valorem_ (property) taxes, they may be repaid with project revenues or any unrestricted income of the government entity, such as sales taxes, license fees, income taxes, and other fees. In most areas of the country, the use of a G.O. debt is contingent upon approval of local voters in a special referendum. Local governments have used G.O. bonds to finance such services as bridges, sewers, airports, stadiums, and housing projects.

G.O. bonds are generally considered the most secure form of municipal tax-exempt bonds which is usually reflected in their relatively low interest cost. Additionally, a G.O. bond issue is the simplest and most readily marketable form of debt to issue since many of the complexities and potential delays associated with alternative methods of financing are not associated with G.O. bonds. One disadvantage, however, is that the government's borrowing capacity for other critical community projects may be restricted if such a pledge is undertaken for its waste-to-energy project. The largest G.O. bond issue for a waste-to-energy project was undertaken by the City of Honolulu for their $145,000,000 refuse derived fuel system.

Project Revenue Bonds

Revenue bonds are limited obligations of a community, only secured by the revenues expected to be generated from the operation of the facility. This includes all solid waste user fees, revenues from the sale of power and recovered materials, as well as investment income. Typically, the issuing government pledges in a bond rate covenant to fix and collect rates and charges for services rendered by the waste-to-energy facility sufficient at all times to pay operating expenses, bond principal, and interest. Such bonds are secured by trust indentures which control the use of the facility and sources of its revenues.

Revenue bonds can generally be scheduled with maturities ranging up to 30 years. Revenue derived from the operation of the facility may be the sole security for the bonds. More commonly, additional security in the form of revenue from other sources must be pledged. In order to issue revenue bonds, the government's political body must adopt a bond resolution specifying the application of bond proceeds to construction of the facility, creating a lien on revenues of the facility, establishing a flow of funds and, in general, setting forth in detail the rights of the bondholders and obligations of the issuing government.

The rate of interest charged on such bonds, although higher than G.O. bonds, is thus based upon potential investors' or bond rating agency (i.e., Standard and Poor's and Moody's) perceptions of the financial viability of the project;, the contractual arrangements among the contracting parties; and perceived value of the local government's back-up pledge of revenues. Such perceived risks require that all project revenue bond financing must have debt service reserve accounts to protect against unforeseen revenue shortfalls. For example, the local government may be required to maintain a debt service coverage ratio of 1.0 to 1.5.

Industrial development revenue bonds (IDBs), a distinct form of revenue bonds, have been the most widely used form of financing for waste-to-energy facilities. These bonds are issued by local government to finance the construction of such plants which are then leased or sold to a private corporation.

Federal tax law, which is exceedingly complex and technical in nature, affects the issuance of IDBs to finance the construction of waste-to-energy facilities. The recently enacted Tax Reform Act of 1986 has dramatically altered the use of IDBs to finance such projects. The following is a brief discussion of this tax area. The reader is cautioned that the community's tax counsel should be consulted for specific and detailed recommendations.

With respect to waste-to energy projects, bonds issued by communities to finance such projects are generally regarded to be IDBs if one of the following is true about a project:

- o Energy (steam, electricity, or recovered materials) is sold to a taxable entity such as an investor-owned utility or private company; or

- o A private contractor operates the project with a contract length in excess of five years; or

- o A private contractor bears a "risk of loss" for situations beyond its control; or

- o Solid waste is delivered by private haulers (not agents of the government) and their revenues account for more than 25 percent of the total project revenues.

IDBs for publicly-owned, waste-to-energy projects are currently tax-exempt if 95 percent of the bond proceeds are used for the solid waste disposal portion of the project. Under the so-called "95/5 rule", expenses for the construction or installation of equipment related to the sale of byproducts from the plant (steam, electricity, or recovered materials) are non-exempt. Therefore, these expenses, which are typically ten to 15 percent of the total project cost, must be financed with taxable debt.

The Tax Reform Act of 1986 reduced the tax-exempt IDB allocation for privately-owned, waste-to-energy projects. In contrast to public waste-to-energy projects, privately-owned projects must compete against all types of industrial development projects for a share of a state's allocation during the year. This allocation has been set at $75 per capita, or $250 million per state, until 1987. After that year, the limit is $50 per capita, or $150 million per state.

Grant Funds, Loan Guarantees, and Entitlements

Federal programs were once available to finance the implementation and construction of waste-to-energy projects. These funding sources are either no longer available or extremely limited at the present time. Several states such as New York, Washington, and Florida have established either grant or loan guarantee programs.

Private Equity

Equity capital contributions from the facility developer or third parties, such as banks, insurance companies, corporations, and private investors, can reduce the overall capital needs for a particular project. In return for his equity investment of some 15 to 25 percent, for example, the private equity contributor becomes the owner of the facility, and is entitled for tax benefits, if any, under existing law, and any accrued revenues from the project (e.g., tipping fees, energy and materials recovery revenues, etc.). This private owner also becomes the "deep pocket" for repayment of the project bonds. A great many waste-to-energy projects were financed in recent years utilizing contributions of equity capital from private investors. Due to the loss of favorable tax benefits to private investors under the Tax Reform Act of 1986, this level of private equity contributions in the waste-to-energy projects may be reduced in the years ahead. Additionally, the ten percent investment tax credit has now been eliminated, and depreciation for plant equipment has been extended from five to ten years. For those projects whose tax benefits were grandfathered under the liberal transition rules under this Act, equity contributions may continue to be available from private investors. The level of such equity contributions will depend in part upon the value of the tax benefits associated with the equity investment and the relative rate of return of these equity investments in waste-to-energy projects as compared to alternative investments.

KEY PARTICIPANTS IN RESOURCE RECOVERY FINANCINGS

The implementation of a complex public works project such as a waste-to-energy project requires the specialized technical, financial, and contracting skills that are usually not found in local government. Consequently, public bodies contemplating these projects commonly acquire professional services to supplement in-house capabilities in such areas as engineering, finance, insurance, environmental science, and law. The financing of a waste-to-energy project involves these key participants beyond the principals to the transaction. The roles and responsibilities of each of these participants are discussed in the paragraphs below.

Bond Counsel

The role of the bond counsel in waste-to-energy and other public financings is quite different from that of counsel to types of corporate financings. In public financings, the legal counsel essentially does not serve as attorney for any particular party, but acts as a sort of special counsel for the transaction itself. That is, the main function of the bond counsel is to render an opinion regarding the securities to be issued by the public body to pay for the construction of the facility.

As expected, the contents of this opinion vary from project to project. Essentially, however, the bond counsel is responsible for reviewing the bonds themselves as well as the official governmental documents, certificates, legal agreements, laws, decisions, and rulings authorizing their issue to ensure that the bonds were issued in a legal manner and constitute valid and binding obligations of the public body. If the bonds to be issued are to be secured by project revenues, the bond counsel's opinion will include a statement regarding the legality of pledging such revenue sources according to federal, state, and local regulations. Additionally, where the interest on the bonds are to be exempt from taxation, the bond counsel will be called upon to render a legal opinion as to the tax-exempt status of the interest on these bonds.

In addition to the duties outlined above, the bond counsel retained by the public body may be called upon to prepare the necessary documents, such as the trust indenture and bond resolution, in connection with bond issuance and to advise on other matters related to the financing of the project.

Independent Consulting Engineer

When a project is ready to be financed, an independent consulting engineer is hired by either the issuer or the underwriter of the bonds to prepare a feasibility report on the project. The independent consulting engineer is charged with reviewing the preliminary engineering plans, specifications, and cost estimates, which are available at that time, for construction and operation of the waste-to-energy facility. Generally, the independent consulting engineer, whose report is included in the official statement, must make a determination about the project in the following areas:

o Is the project technically feasible?;

o Are the estimates of the amounts of solid waste available for combustion and processing reasonable?;

o Is the construction cost for the facility appear to be reasonable and consistent with costs of similar facilities?;

o Were the construction and operating cost estimates prepared using sound engineering estimating methods?;

o Are the estimates of the amounts of revenues and expenses produced by the facility reasonable?; and

o Are the terms of the construction and operation agreements in the best interest of the project?

Investment Banker

The investment banker or bond underwriter plays a principal role in the marketing of the bond issue. When such bonds are to be issued, the investment banking firm, either alone or with other partners, makes a proposal to the municipality for the purchase of its bonds, which are in turn sold by the investment banker(s) to investors. The investment banker's profit is derived from the spread between the price at which the bonds are purchased from the municipality and their ultimate resale to investors.

Out of this spread, these firms must pay certain expenses associated with the bond issue such as: the fee of the underwriters' counsel, out-of pocket expenses, and expenses incurred in connection with the delivery of the bonds.

The investment banker serves an important function for the municipality during the period of time in which bonds are being issued by assuming the risks of adverse price fluctuations. It is possible that the market price of these securities will decline before all of the issue is sold. As part of its fee, the investment banker underwrites this risk and insures that the municipality will receive the amount of capital funds necessary for the project.

Financial Advisor

Although financial advisors can be hired by any of the parties in a waste-to-energy project, they are usually employed by the municipality to review the financing plan, and to advise the community on bond marketability and the expected interest rate based on this financing plan.

Trustee

A trustee is a trust company or bank with trust powers, authorized to do business within the particular state, who performs the functions defined in the indenture. The Indenture of Trust or Trust Indenture is a legally binding agreement between the municipality and the trustee in which the municipality pledges its interest in the project as security for the payment of principal, premium, if any, and interest on the bonds issued for construction and operation of the waste-to-energy project.

The trustee may be responsible for the following tasks: dispersing the construction fund; directing the flow of revenues from the project; investment of monies in various funds as set forth in the indenture; monitoring the levels of these different funds; paying the principal and interest payments on these bonds; and for seeing that the bondholder's interests are protected during the terms of the bonds.

Rating Agencies

Bonds are assigned quality ratings that reflect their relative investment qualities. There are two major bond rating agencies (Moody's Investors Service and Standard and Poor's Corporation). Bonds with a rating of triple A (AAA) have the highest rating and are felt to be extremely safe, while double A (AA) and single A (A) bonds are often held in conservative portfolios. Triple B (BBB) bonds are considered to be investment grade bonds, and, as such, are permitted by law to be held by banks and other institutions.

Such bond ratings are very important to both the issuer and the investor for many reasons. First, a bond's rating is an indicator of its risk. Thus, a bond with a high rating usually results in a lower interest rate than one with a low rating. Second, most bonds are purchased by institutions, which are restricted by law in many areas of the country to investment grade issues. Consequently, if a bond receives a rating below triple B, the municipality may have a difficult time selling these bonds.

Prior to the bond offering, the municipality or its representative presents to the rating agencies all the pertinent documents related to the project's financing such as the indenture, consulting engineer's feasibility report, construction and operation contracts, and municipal bond insurance policy, if any. Analysis of these documents by the rating agencies forms the basis of their initial or conditional rating. There is no guarantee that such ratings will continue for a given period of time or that they will not be lowered or withdrawn entirely by the rating agencies if in their judgment circumstances so warrant.

STEPS IN BRINGING THE BOND ISSUE TO MARKET

Having noted the principal participants responsible for financing a waste-to-energy facility, discussed in the paragraphs below is a simplified listing of the major general steps and documents necessary to bring a waste-to-energy bond issue to market.

Step 1. Adoption of a Bond Resolution and Trust Indenture

Revenue bonds issued by a municipality are secured by contracts and legal documents which must be enacted by the governing body of the community. A bond resolution sets forth the maximum size of the issue; officially authorizes the various financing techniques which may be required to successfully market the bond issue (e.g., municipal bond insurance, letters of credit, discounts, call provisions, etc.); authorizes the execution of an Indenture of Trust between the municipality and a trustee; outlines the procedures for obtaining interim borrowing (e.g., issuance of bond anticipation notes); and authorizes the commencement of bond validation proceedings.

As noted previously, a Trust Indenture is a contract between a municipality and a trustee which contains provisions establishing adequate financial reserves for protecting the bondholders should problems develop in the plant. It also specifies how the proceeds of the bond issue and the revenues generated by the project during its operation will be invested; how obligations to the contractor and bondholders will be paid; how much the trustee will be compensated; the circumstances under which the trustee can be removed; procedures for the modification and amendment to the indenture; and the discharge of the Indenture. To carry out these major tasks, the Trust Indenture will provide for the creation of special funds such as: debt service; revenues; operating and maintenance; reserve and contingency; subordinated debt service; state loan repayment; and renewal, and replacement.

Step 2. Validate the Bonds

Local courts in most localities have jurisdiction to determine the validation of bonds. Any government entity may file a complaint in such courts where the proceeds of the bond issue are to be expended to determine its authority to incur bonded debt. This complaint is required to include information relative to the ordinance, resolution, or other legal proceeding authorizing the issue; the size of the issue and the amount of interest on each bond; and other pertinent matters for the court to weigh in its deliberations. The date for the hearing is required to be published well in advance of the hearing.

By requiring this legal notice, all citizens or taxpayers of the local area are made parties to the proceeding as if they were personally served. Consequently, any citizen or taxpayer can intervene in the validation proceeding. If the final judgement of the court validates the bonds, and no appeal is taken within the prescribed time, then the judgement is forever conclusive, and the validity of such bonds can never be called into question in any court by any person or party. However, if any person is dissatisfied with this final judgement then he can usually appeal to the state's highest court within the time and manner prescribed in the state's appeallate rules.

Step 3. Preparation of a Preliminary Official Statement

An official statement is a document issued by the investment bankers or underwriters to the community in connection with the issuance of the bonds. The official statement for a waste-to-energy bond issue will typically contain an introductory statement describing key financial and other features of the issue, and is followed by sections which summarize the pertinent financing documents and agreements, including the bond resolution, trust indenture, construction and operation contracts, the energy market contract (for energy produced and materials, if any, recovered by the facility), lease, or other contracts with third parties. The official statement will have certain appendices, such as the independent consulting engineer's report, audited financial statements, and the opinion of the bond counsel as to the tax-exempt status of the bonds.

Generally, the official statement is preceded by a preliminary official statement which is issued primarily for marketing purposes and before the pricing of the bonds. Such documents are used by many investment banking houses to establish pre-sale interest in the issue and identify potential buyers among insurance companies, banks, investment bankers, municipal funds, pension funds, and others who may be interested in purchasing the local government's bonds. During this period the underwriter's sales force makes contact with these prospective institutional and retail investors, which may include meetings and seminars in the major regional and national financial centers of the United States, to determine the level of interest in the bond issue and potential coupon rates that may be necessary to attract the investment community in the issue. Although the final official statement normally follows the preliminary official statement except for minor changes, the later document has a statement printed in red along the left margin of the front cover admonishing the purchasers of the issue that they should only rely on statements in the final official statement. This document is commonly referred to as the "Red Herring".

Step 4. Meetings With Rating Agencies and Bond Insurance Firms

Bond ratings are key to the successful financing of an waste-to-energy project. As indicated previously, both rating agencies realize that each waste-to-energy facility is a unique project, with unique security features and project risks. Consequently, the analysis of each project is tailored to that project's particular legal and economic structure. The local government's investment bankers will at this point assist in providing all information necessary to these various rating agencies so that the bond issue can be properly evaluated.

Commonly, this information is presented through visits to the rating agencies by project personnel. This allows the municipality to more fully portray the need for its project and outline the managerial and financial strengths of the municipality. In developing these presentations to the rating agencies and bond insurance firms, communities should have access to the municipal bond research staff of the underwriters since they often have prior rating and insurance placement experience gained on similar waste-to-energy projects.

Step 5. Blue Sky and Legal Investment Surveys

"Blue-sky" laws are those state laws regulating the sales of securities. Such laws were passed by most states in the early part of this century to prevent unscrupulous security salesmen from operating in their cities and communities. The term "blue-sky laws" can be traced to a member of the Kansas legislature who remarked that such promoters would sell stock in the "blue-sky" itself.

Today, most states and U.S. possessions have enacted "blue-sky" and investment laws. The "blue-sky" laws typically set forth the procedure under which new security issues and dealers or brokers of such securities are registered with the state. Many states also have enacted investment laws which regulate the security portfolios of financial institutions such as savings banks, insurance companies, trustees, and municipal corporations. Since these regulations may affect the marketing of a bond issue, it is important that a survey of the different state laws be done before the bonds are issued. This survey is usually completed by the underwriter's counsel at the time the preliminary official statements are mailed to prospective investors.

Step 6. Establishing the Final Pricing of the Issue

Prior to the submittal of a bond purchase contract to the municipality, the investment bankers arrive at a final offering price of the bonds based upon the results of their pre-sale distribution efforts. At the several pricing meetings held during this period, members of the investment banking syndicate, which were selected by the municipality for this particular issue, compare their expectations about capital market trends, the supply of competing bond issues on the financing calendar, secondary market activity, the size of the bond issue, and the buying patterns of investors.

Step 7. Submission of the Purchase Contract to the Issuer

Commonly, the timing of the purchase proposal is usually determined by the investment banking syndicate. The investment bankers will negotiate the purchase of the bonds required for the project soon after a full-service contractor has been selected and all necessary contracts have been negotiated and executed. At the time this proposal is submitted to the government entity, the investment bankers must provide its financial advisor with interest rates, spread, management fees and other information to assure the community that the bonds are being sold at the best possible price given the municipal bond market at that time.

Based on an analysis of this information, the municipality can either accept or reject this bond purchase proposal. If this proposal is rejected, the municipality and the investment bankers commonly continue to negotiate in good-faith until an agreement is reached as to the purchase price, interest rates, and other terms or conditions of the sale of the bonds. At such time the community and the investment banking syndicate reaches agreement on these issues, a purchase contract is signed by both parties binding the investment bankers to purchase the bonds at a stated price once the par amount of the bonds are delivered to them by the municipality.

The signing of the purchase proposal sets in motion several important activities. Commonly, it is at this time that the final official statement is printed and then distributed by the managing underwriter to other members of the underwriting syndicate, institutional and retail investors, and the municipality. Also, the managing underwriter coordinates the allocation of the bonds among the syndicate members and actively monitors on a day-to-day basis the actual selling of the bonds to investors.

Step 8. Bond Issue Closing

On the day of the closing, the municipality delivers to the investment bankers its legally executed and binding definitive bonds. Documents relative to the bond issue are again reviewed by the local government's bond counsel and the underwriter's counsel to be certain everything is in order. Simultaneously with the delivery by the municipality of its duly executed bonds, the investment bankers deposit with the trustee the full purchase price of the bond issue, as specified in the purchase proposal and the trust indenture.

Step 9. Post Sale Activities

After the bond issue has been sold and the day of the closing has passed, many investment bankers will continue to play an active role in the local government's bond issue. Some firms will help maintain a secondary market in these bonds even under unfavorable market conditions. Generally, it is the practice of most firms in the industry to continue to assist their municipal clients in maintaining regular contacts between the municipality and the rating agencies to keep the agencies informed as to the financial situation of the local government's waste-to-energy project.

REFERENCES

1. Bolczak Richard and Robert E. Zier, Financing Options and Alternatives. *Solid Wastes Management* September: 185-189 (1982).

2. Franklin, Wendy and Stephen Howard, Tax Reform's Impact, *Waste Age*, November: 49-54 (1986).

3. Hillsborough County Project Team, *Decision Paper on Ownership and Financing of the Hillsborough County, Florida Solid Waste Energy Recovery Facility*, Tampa: Hillsborough County, Florida (1983), pp 1-16.

4. Nemeth, Diane M., *Resource Recovery Option in Solid Waste Management: A Review Guide For Public Officials*, Chicago: American Public Works Association (1981).

5. New York State, *Resource Recovery Procurement and Financing: A Guidebook for Community Action*, Albany: New York State Department of Environmental Regulation (1980), pp. 6-1 - 6-7.

6. Piliero, Mark R., *Resource Recovery Revenue Bonds*, New York: E.F. Hutton and Company, Inc. (1986).

7. Rappaport, Stephen P., *An Invesor's Guide to Resource Recovery Bonds*, New York: Prudential-Bache Securities (1985).

8. Shearson Lehman Brothers, Inc., *Impact of Tax Reform Legislation on Resource Recovery Projects*, New

York: Shearson Lehman Brothers, Inc. (1986).

9. Smith, Cyril V., and Johm B. Pirog, The Role of the Counsel to the Underwriters, In: *The Municipal Bond Handbook*, Homewood, Illinois: Dow-Jones Irwin (1983), p.244.

10. U.S. Environmental Protection Agency, *Resource Recovery Plant Implementation: Financing*, Washington: U.S. Environmental Protection Agency (1975), pp. 4-8.

Index

Aesthetics - 95, 165
Air pollution control
   baghouse - 38
   electrostatic precipitator - 38
   scrubbers - 38
   standards - 93, 124-133
Air quality modeling - 94, 103
Akron, Ohio - 52, 76-78
Albany, New York - 52, 57
Aluminum - 120
Ames, Iowa - 3, 52
Anaerobic digestion - 56
Andco-Thorax - 58
Ash
   residue - 19, 39
   quench bath - 43
Attainment area - 128
Auburn, Maine - 45
Avoided costs - 115

Bern, Switzerland - 34
BACT - 131
Baghouse - 38
Baltimore, Maryland - 26, 41
Battelle Columbus Laboratories - 61
Black Clawson - 47
Boilers
   dedicated - 51
   economizer - 38
   fluidized bed - 51, 54-56
   refractory lined - 34, 38
   waterwall - 34, 38
Bonds
   bond counsel - 184-186
   closing - 196
   general obligation - 178-179
   industrial development - 177, 180-186
   insurance - 25
   payment and performance - 25
   purchase contract - 195
   rating - 188
   resolution - 189
   validation - 190-191
Bottom ash - 39
Buffer zones - 89, 95

Calorimeter - 71
Capacity credit - 115
Carbon monoxide - 127 (see also Clean Air Act, criteria pollutants)
Chicago, Illinois - 41
Chlorinated hydrocarbons - 139
Clean Air Act (see also Air pollution control standards)
   criteria pollutants - 126

Clean Water Act - 125
Coal - 50
Cockeysville, Maryland - 52
Collegeville, Minnesota - 45
Collier County, Florida - 55
Columbus, Ohio - 52
Commerce, California - 26
Committees
    citizen review - 28
Composting - 56–58
Contractual flow control - 80–81
Corps of Engineers - 125, 135
Costs - 25
Crane - 35
Cuba, New York - 45

Dade County, Florida - 48, 52
Denmark - 4
Dioxin - 139–141
District heating - 46
Dredge and fill - 135 (see also Corps of Engineers)
Duluth, Minnesota - 52, 55

Electrical transmission - 92
Energy efficiency - 58–60
Energy sales
    avoided capacity - 115
    avoided energy - 115
    electricity - 113
    importance of markets - 9
    steam - 111–113
Engineering consultant - 13
Erie, Pennsylvania - 55

Feasibility study - 20
Federal Aviation Administration - 96, 135–136
Federal Energy Regulatory Commission - 115
Financial advisor - 13
Financing
    blue sky opinion - 193–194
    bond counsel - 185–186
    consulting engineer - 185–186
    grants - 182
    investment banker - 186
    official statements - 25, 191

    private equity - 183
    "red herring" - 192
    trustee - 187–188
Flail mills - 48
Flow control - 74–82
Fluidized bed incineration - 51, 54–56
Fly ash - 39
Food wastes - 70
Franklin, Ohio - 47, 55
Full service vendor - 150

Gallatin, Tennessee - 41
General obligation bonds - 178–179
Glass - 120
Glen Cove, New York - 41
Go/no-go decision - 19
Grates - 120

Hamburg, Germany - 33
Hampton, Virginia - 41
Hart Report - 142
Hazardous air pollutants - 128
Hempstead, New York - 47
Hennepin County, Minnesota - 26
Hillsborough County, Florida - 12, 26, 157
Hydropulper - 47–48

Insurance advisor - 13

Japan - 2, 39

Keystone siting process - 6

LAER - 133
Lakeland, Florida - 53
Lawrence, Massachusetts - 26, 53
Lead - 127 (see also Clean Air Act, criteria pollutants)
Legal counsel - 14
Lowest achievable emission requirement - 133

Madison, Wisconsin - 53
Marion County, Oregon - 41
Mass burning incinerators - 5, 33–42

Index 201

Materials markets - 9, 33-42
Metals
  ferrous - 70, 119
  non-ferrous - 70, 119-120
Modular combustors - 5, 42
Monsanto - 58

NAAQS - 126-128
Nashville, Tennessee - 41
New Hanover, North Carolina - 41
NEPA - 125
NESHAP - 128
Niagara Falls, New York - 53
NIMBY - 2
Nitrogen dioxide - 127 (see also Clean Air Act, criteria pollutants)
Nonattainment area - 128
Norfolk, Virginia - 26, 40-41
North Andover, Massachusetts - 47, 81
North Little Rock, Arkansas - 43, 45
NPDES - 134-135
NRRA - 40

Occidental - 58
Olmsted County, Minnesota - 26
Operators
  full service - 19, 23-24
Osceola, Arkansas - 45
Oswego, New York - 45
Ownership
  private - 23, 174-175
  public - 23, 174-175
Ozone - 127 (see also Clean Air Act, criteria pollutants)

Paper products - 70, 121
Particulates - 127, 130 (see also Clean Air Act, criteria pollutants)
Pascagula, Mississippi - 45
Permitting - 96, 123
Pinellas County, Florida - 3, 26, 41
Pittsfield, Massachusetts - 45
Plastics - 70, 121
Pompano Beach, Florida - 56

Portsmouth, New Hampshire - 45
Prepared fuels - 5 (see also RDF)
Procurement
  A/E - 18, 24, 148-149
  "chute to stack" - 18
  competitive negotiations - 154
  RFP - 22, 155, 157-169
  RFQ - 22, 155
  simultaneous negotiations - 152-154
  sole source - 156-157
  turnkey - 15, 149-150
Project
  cost - 60
  elements - 7
  manager - 12
  numbers - 5
  team - 10-14
Public information programs - 25-28
PURPA - 113-114 (see also Federal Energy Regulatory Commission)
Pyrolysis - 51

Qualifying facility - 114 (see also Federal Energy Regulatory Commission)

RDF
  densified RDF - 48
  dry processing - 48-54
  fluff RDF - 48
  powdered RDF - 48
  wet processing - 47-48
Recycling - 65, 80
Revenue bonds - 179-182
Risks
  construction - 18
  energy - 17
  force majeure - 14
  legal - 17
  operation - 19
  waste stream - 16
Rotary kiln - 37
Rubber and leather - 70

Salem, Virginia - 45

Sanitary landfilling - 1
Saugus, Massachusetts - 3, 26, 41
Scales - 71
Sherman Anti-Trust Act - 77–78
Siting
   candidate sites - 104–106
   comparative costs - 107
   constraint mapping - 98–100
   environmental considerations - 93–95
   rating - 101–103
   selection - 85
   social considerations - 95
   technical considerations - 88–93
Sizes - 4–5, 39
Sludge incineration - 55
Solid waste
   commercial - 66
   composition - 67–69
   heating value - 69–71
   industrial - 67
   quantities - 71
   residential - 66
   special - 67
St. Louis, Missouri - 50
Standard contract offers - 116
Starved air incineration - 42
Statewide avoided unit - 117
Storage pit - 35, 89
Sulfur dioxide - 127, 130 (see also Clean Air Act, criteria pollutants)
Sweden - 4
Switzerland - 4

Tampa, Florida - 26, 41
Tax code - 3, 176, 181
Technology
   European - 39
   evaluation - 58
   selection - 31–32
Textiles - 70
Tipping floor - 35
Trommels - 48
Tulsa, Oklahoma - 41
Tuscaloosa, Alabama - 45

Union Carbide - 58
Utilities
   potable water - 93
   sanitary disposal - 93

Waste flow control - 16, 74–83
Water quality - 95, 133
Westchester County, New York - 41
Wet-pulped RDF - 47–48
Wilmington, Delaware - 53, 57
Windham, Connecticut - 45
Wood - 70

Yard wastes - 70